ADVANCED GARMENT CONSTRUCTION GUIDE

ADVANCED GARMENT CONSTRUCTION GUIDE

Dr. M. Sumithra

WOODHEAD PUBLISHING INDIA PVT LTD

New Delhi

Published by Woodhead Publishing India Pvt. Ltd.
Woodhead Publishing India Pvt. Ltd.,
303, Vardaan House, 7/28, Ansari Road,
Daryaganj, New Delhi - 110002, India
www.woodheadpublishingindia.com

First published 2020, Woodhead Publishing India Pvt. Ltd.
© Woodhead Publishing India Pvt. Ltd., 2020

Woodhead Publishing India Pvt. Ltd. ISBN: 978-81-936446-4-5
Woodhead Publishing India Pvt. Ltd. e-ISBN: 978-81-936446-5-2

Typeset by Allen Smalley, Chennai

Digitally Printed and bound by Replika Press Pvt. Ltd.

Contents

Preface

Advanced garment construction book will teach how to design and sew garments that flatter their figure and reflect their personal style. Garment making is one of the basic content of fashion designing. Fashion designers work in a number of ways in designing clothing and accessories, Fashion designers attempt to design clothes which are functional as well as aesthetically pleasing. Fashions always change, but the principles of garment designing and pattern drafting and the techniques of construction remain basically the same. With the help of principles, hundreds of new styles and designs can be created according to change in fashions.

This book presents practical working directions for the construction of clothing. This is a perfect book of those who want to make their own clothing for style and fit .The entire process of Garment construction, including a section on the Measurement, Material requirement, Information on working with patterns and construction details with illustration and cost calculation. Chapters covered as Recent fashion wear, Specialized work wear, Fashion show garments, Need based garment and Garment accessory making.

Garment making is thus a technical accomplishment that requires knowledge of fabrics, principles of clothing construction and skills involved in it. This depends on the ability to select the correct color, design, fabric and accessories to suit an individual occasion. A garment that is made will be attractive if it fits well and proper attention is paid to its finer details. This book serves as a guide for beginners and students of fashion designing and apparel manufactures.

PART I

Advanced garment construction

1.1   Introduction

Clothing is one among the most important three basic needs in every human life. It protects our body from various climates and gives us good appearance. A dress satisfies social and emotional needs but it is worn on body and it becomes part of our physical being. Clothing refers to a covering of all human societies. Clothing performs a range of social and cultural functions, such as individual, sexual differentiation, occupational and social status. Clothing reflect standards of modesty, status, gender, religion and expression of personal taste.

Construction of a garment is a beautiful art, which requires skill of sewing which is essential to convert the design on paper in to garment. Garment construction has both technical and design issues, the designer can choose where to construct lines like pockets, collars, plackets, sleeves and how to finish edges and how to produce volume and structure in order to create a good look and experience for the wearer. Garments we design ourselves appeal to the emotions and attractive to the eye. Unexpected forms and shapes can be converted in to flat patterns and ultimately in to garments. Patterns are like documents that describe a garment, conveying its more structure. The success of a design led approach relies upon your wearer engaging with the garment during the use, whether this involves participating with all of the features offered by a garment.

A garment is constructed by cutting the fabric in to parts according to a pattern which fits the human form then the different parts are joined together by sewing. Garment making requires knowledge of fabrics, principles of clothing construction and skills involved in it. This depends on the ability to select the correct fabric, accessories, design and colour to suit an individual occasion. Garment construction is usually guided by a pattern. A pattern can be simple, some patterns are nothing more than a mathematical formula that the sewer calculates based on the intended measurements. Once calculated, the sewer has the measurements needed to cut the cloth and sew the garment together. Pattern making is a complex task as no two human beings are identical.

Pattern making have generally a good system to be fully satisfy the needs of human figures. The human proportionate systems work on the principles that the human body length is divided in to eight heads and the girth measurements are in equal proportion in to each other. Pattern making systems are mainly dependent and are influenced by accepted trend of their period. A garment that is made will be more attractive if it fits well and proper attention is paid to its finer details. Pattern making is for modifying and shaping a flat piece of fabric to conform to one or more curves of the human figure. Pattern making is a bridge function between production and design. A sketch can be turned into a garment through a pattern which interprets the design in the form of the garment components.

Sewing is a creative and interesting skill. A good sewing machine is required to obtain quality products, depending up on the ability and requirements of the person.

Before beginning it is necessary to consider the following points.

1. The measurements required for the garment.
2. The different parts of which it is composed.
3. Suitable material with regard to cutting advantage, cost, qualities of wearing, ease for laundering and repairing.
4. The material preparation for drafting and cutting.
5. The drafting of the pattern.
6. The laying the pattern above the material.
7. The cutting out.
8. The construction of the garment.
9. The trimming.
10. The pressing.
11. The cost.

1.2   Cutting and making of garments

General rule for the construction of garments are as follows:

1. Taking of body measurements.
2. The drafting and placing of the patterns.
3. The cutting of the pattern according to the garment.
4. The joining of the different parts.
5. The fitting and necessary alterations.
6. The stitching and finishing of the garment.
7. The application of trimmings, if these are used.
8. The pressing.

1.2.1   Taking of body measurements

Taking of measurement is very important step, because the fit and appearance of garment depend upon the accuracy of the measurement taken. Use a good quality measuring tape for taking body measurements, which is steady and will not stretch and it should not be too stiff and it should be pliable. Take all measurements closely but not tightly. All vertical measurements should be taken on one side only. All horizontal measurements should be taken with the tape measure parallel to the floor. The measurements needed for the garment construction depends on style, type, age and sex of the wearer.

- The measuring tools are
- Measuring tape
- L-square
- Rulers
- Yardstick (or) Meter scale
- Skirt marker
- Hem gauge
- Hem marker.

The **ladies' measurements** are
Bust, waist, hip, back width, armscye depth, distance between bust point, lower arm circumference, wrist, back waist length, neck, shoulder, front waist length, shoulder to bust, full-sleeve length, short-sleeve length, upper-arm circumference, elbow circumferences, wrist and sleeve length. Waist to hip, waist to ground, choli length, pant top length, kurta length, maxi dress length, maxi skirt length and middy skirt length.

The **children's measurements** are
Chest, waist, hip, neck, back width, back waist length, armscye depth, short-sleeve length, lower-arm, full-sleeve length, cuff length, shirt length, pant length, half-pant length, wrist, waist-to-hip, thigh girth, crotch length, cervical height, waist to ankle, maxi skirt length, short skirt length, frock length and blouse length.

The **men's measurements** are
Chest, waist, seat, neck, back width, cuff length, back waist length, short-sleeve length, full-sleeve length, shirt length, kurta length and pant length.

1.2.2   The drafting and placing of the pattern

Drafting may be defined as a system of drawing patterns on paper with mechanical precision on the basis of body measurements. This is an effective and economical method which can be learned easily.

Pattern making tools used for drafting are as follows:

- Pins
- Pin cushion
- Orange—stick
- Scissors
- Ruler
- Cutting table
- Tailors square
- Measuring tape
- Tailors chalk
- Tracing wheel
- Tracing paper.

Principles for pattern drafting: While drafting patterns, the following principles have to be followed.

1. Select a suitable paper for drafting patterns which should not be too thin.
2. Use suitable instruments like long scale, "L" type scales, Set squares, French curves, etc., for accurate drafting.
3. Use a sharp pencil for fine and neat lines and accurate drafting.
4. Before drafting the patterns, check the measurements clearly and read the procedures and instructions carefully.
5. Make a rough diagram before drafting. This will give an idea for drafting patterns with perfection.
6. Before drafting, we have to decide the amount of ease allowances to be given at the required portions. The prepared patterns must be larger than body measurements to allow for freedom of movement, ease of action and comfort in wearing. For that ease allowance are given along with the body measurements for free body movements.
7. Seam allowance should be decided first before drafting the patterns. According to the seam and stitch types, we have to give seam allowance at the outlines of the patterns.
8. If a pattern has symmetric design where the right and left sides are alike, we may make only the part of the pattern. For asymmetric designed patterns, full pattern must be drafted.
9. Check the draft before cutting the patterns.
10. After drafting the patterns, the following details and information should be recorded and marked clearly on the patterns.
- Name of the style (e.g., Men's shirt, Skirt, etc.)

- Name of each piece of pattern (e.g., Front, Sleeve, etc.)
- Size of the garment (e.g., M, L, Chest measurements, Hip measurements, etc.)
- Number of pieces to be cut with each pattern piece, (e.g., for shirt, we have to cut 2 fronts, 1 back, 2 yokes, 2 sleeves, 2 collars and 2 collar bands)
- The cutting line should be clearly marked. After that the seam line should be marked using dotted lines.
- Grain line should be drawn on all the pattern pieces. This line indicates that the patterns should be kept on the cloth in such a way that the line is parallel to the length of the cloth or the selvedge.
- The position of buttons and button holes should be marked.
- Fold line, should be clearly shown. Bottom hem line should be marked to show where the material is to be folded.

Centre front and centre back line should be marked. Inward and outward notches should be marked to identify the position of pleat, dart, hemline, etc.

Pattern: Pattern is developed from the block that includes all the information needed for cutting and production of the garment including seam allowance.

Collar: 2 piece collar–¼″, all side Chinese collar–¼″. All side arm hole–½″ on the curve, Placket: ¼″–½″, Hemming: ½″–2″, Sleeve: ½″ on the curve and 1″ on the under arm seam. Sleeve hem: Without cuff: ¾″–1″, with cuff: ¼″ Cuff: ¼″, Pocket body: ½″ Upper: ¾″, Flap: ½″, Side seam: ½″–1″, Yoke: ¼″ Patch: ¼″.

For the bottom waist band: ¼″, Fly: ¼″, Side seam: ½″–1″, In seam: ½″, Hemming: ½″–1″, Pocket: ¼″–½″, Patch: ½″, Divider flap: 1″.

Note:

- To join two pieces, always keep ½″ seam allowance.
- Keep ¼″ extra allowance for over-lock.
- For finishing the side seam keep ¾″ seam allowance.

Seam allowances: The amount of seam allowance required for each seam line may vary depending on end use. The measurements are as follows:

- ¼″ for curve shapes.
- ½″ for armhole, neckline, waist line, style line.
- 1″ for side seam, centre line, shoulder, plackets.

After paper pattern is drafted, cut out the pattern pieces very carefully and cut as many parts are required to make the cloth in to garment. Mark all pattern details on the right side of the pattern. The material must be unfolded and stretched evenly without any wrinkles on a large table. If the material is not even, it is well to pull the corner. Place all the large pieces first and as economically as possible. The larger pieces should always be cut first and should be placed with the widest part near the cut edges of the cloth. If more pieces is required, be sure to place all the parts of the pattern on the cloth before beginning to cut them out. This may save the material. The parts of the pattern be placed on the fabric with the construction details running in the same direction as the warp and the weft. The basic terms should be understood before proceeding to prepare the fabric.

Grain—The direction of yarn in a fabric is named as grain. On patterns lengthwise grain referred to as straight grain. Woven fabrics are made up of lengthwise and crosswise yarns interlaced at right angles to each other. These yarns are called the lengthwise and crosswise grain. Lengthwise yarns are usually stronger and heavier and stretch less than crosswise yarns.

Selvedge—The finished edge of the fabric which runs lengthwise grain is named as selvedge. In a good quality fabric, the selvedge is about half an inch wide.

On-grain—A fabric in which the crosswise yarns run exactly at right angles to lengthwise yarns and which has right-angled corners is said to be on-grain.

Off-grain—A fabric in which the crosswise yarns are not running exactly at right angles to lengthwise yarns is referred to as off-grain fabric.

1.2.3   The cutting of the pattern according to the garment

The cutting shears well sharpened should be used. Cut with a long stroke to the end of shears to avoid notches in the fabric. The parts of the pattern are pinned exactly and securely to the fabric, trace all around them with a tailors chalk. Then cut the fabric allowing of seams, hems and if fullness is needed to the fabric. Seam allowances ½"–¾", for hems 2"–4" according to the garments. The cutting tools are scissors, button hole scissors, dress makers shears, electric scissors and pinking shears.

1.2.4   The joining of the different parts

Joining the different parts of the garment can be done with pins or basting thread. Check all the corresponding parts meet accurately and also that the connecting points meet perfectly. The sewing tools are sewing machine needles, pins, sewing thread, thimbles, and stiletto, bodkin and embroidery threads.

1.2.5   The fitting and necessary alterations

When all the parts of a garment are pinned together, it is ready for fitting. Mistakes may occur due to careless joining of different parts. The garment is tried on wrong side out. The seam projection on the wrong side makes fitting easier. For a person regularly build, only one side, the right side, is fitted person who have one side difference from the other should always be fitted with right side of the garment out.

Alterations

After the fitting alterations are marked, it is advisable to fit an altered garment before seams are stitched and finished to make sure that every part of the lines is correct.

1.2.6   The stitching and finishing of the garment

No hard rule can be given for the sewing and finishing of garments depends on the type of the garment, material used and type of seams. Careful stitching, good fastening, button hole, neat and clean appearance are essential for the construction of a garment. All pressing must be done, if possible, on the wrong side, but this is especially desirable on coloured cotton cloth, wool or silk.

1.2.7   The application of trimmings, if these are used

Trims enhance the garment appearance. Trims are generally decided by the fashion trend. As they help in creating an effective look with very less effort. Trims such as ribbons, laces, brides, and other narrow fabric, trims are widely used to kids wear, night wear, lingerie, etc. These help in creating a soft look in the garment and without too much effort makes it look good. These have to be chosen that they complement the garment both aesthetically, in terms of decoration, and in terms of ensuring that the garment performs as expected in its end use. There are a large variety of trims and decorative items available in the market. They can be broadly divided into two categories—Functional trims and Decorative trims. Functional trims are those which have a definitive purpose like edge finishes and closures, but they might work as decorative trims, like buttons on the side of the jacket sleeve. The decorative trims are for embellishment only, like laces, ribbons, braids, etc. There are trims that one can buy in the market and there are trims that can be made at home by an individual.

1.2.8   The pressing

Pressing is an up and down motion. Lower the iron, press, and lift and move on to another section of the fabric. Then again lower the iron, press and lift the iron; this is the pressing motion. As ironing has long strokes it is done on flat surfaces. Pressing is done on surfaces using various pads to suit the shape of different pieces and parts of the garment. For the shaped three-dimensional garments, it is advisable to use small pads for pressing. For better finishing, it is advisable to press each and every dart and press-open every seam while stitching. Press-open every enclosed seam before turning the facing back to the underside. It is so much easier to topstitch an edge when it has been pressed flat. It is far easier to insert a zipper after seam allowance has been pressed back first. Do not press over pins or basting as they leave marks. The pressing tools are as follows:

- Iron box
- Steam iron
- Ironing board
- Press cloth
- Sleeve board.

The miscellaneous tools are

- Awl
- Seam ripper
- Loop turner
- Dress form
- Needles
- Thimble
- Bodkin
- Stiletto
- Embroidery thread
- Embroidery scissors.

1.3   Fabric requirement

Purchasing of the fabric for a garment is more important and it requires the person to be expert in pattern making and economical layout. For garment industry it is very important, even minimal saving of 5 cm of fabric in a skirt would results in 50 m being saved in a lot of 1000 skirts. A person buys minimum of 30 cm extra than the required amount does not run short of fabric while cutting. The amount of money spend on extra fabric, which goes waste and is thrown out.

For any garment one needs a minimum of two lengths plus seam allowances. The fabric has two grains—lengthwise grain and crosswise grain. Cut the garment lengths along the lengthwise grain as this is the stronger grain and the fall of the garment would be far better on this lengthwise grain. One is able to cut the garment in less fabric only if the width of the fabric is wide enough to fit two length of the garment in one length of the fabric. The patterns representing all the individual pieces of the garment should be laid out together in such a manner that they fit within the fabric width as closely and efficiently as possible. This minimises the wastage in fabric.

1.3.1   Various possibilities in garment making using different types of Darts, Pleats, Gathers, Tucks, Seams, Necklines, Collars, Sleeves, Plackets, Skirts, Yokes, Pockets, Fastness and Colours.

1.3.1.1   Darts

Darts are small and may be functional, decorative or both. Any dart stitched on right-side of the garment can be said to be decorative dart. Functional darts are for fitting the body curves. The dart is a folded wedge of fabric tapered and stitched down to give shape to S garment. The darts are found primarily more in women's clothing. Eight types of darts are as follows:

1. Plain darts–The plain dart is the most commonly used dart and is usually found around the waist, hip, and bust.

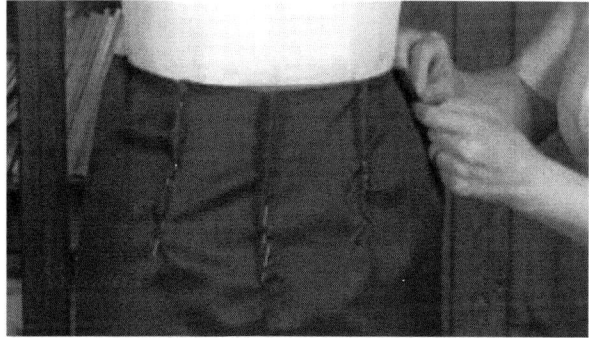

2. Bust darts–Bust darts start from the side seam of a garment and end near the apex of the bust in order to make garment more fitting.

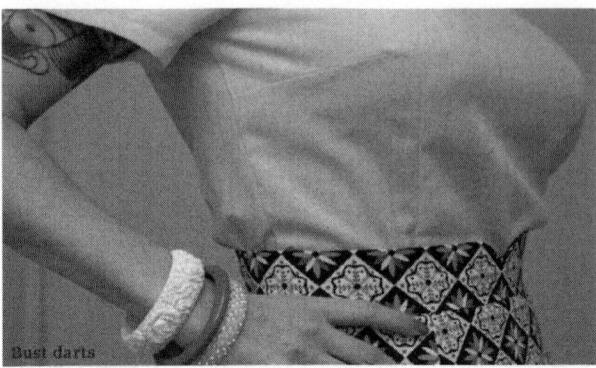

3. French darts–It is a type of elongated bust dart that start at the side seam, down near the waistline and end up near the bust point.

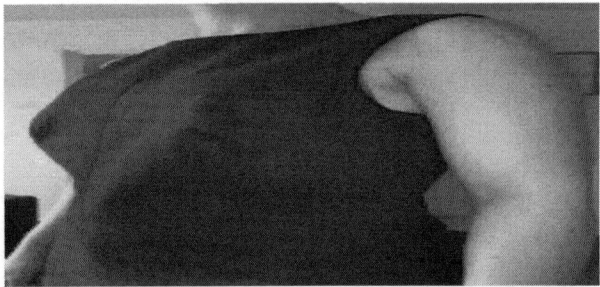

4. Shoulder darts and neckline darts–These darts look feminine.

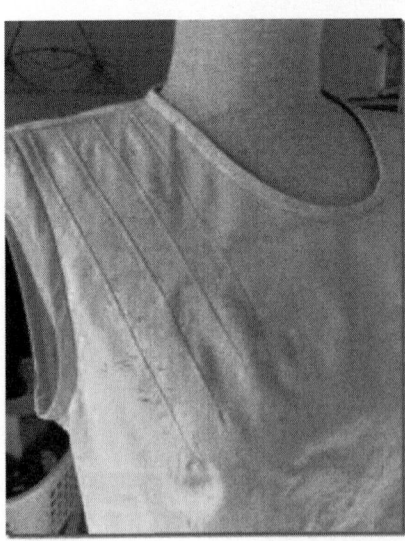

5. Dart tucks–Variation of darts that are constructed like points are left open with pleated appearance. It is mostly used in pants and blouse.

6. Elbow darts–Elbow darts usually keep a slight bend at the elbow and it looks like natural blend. This dart mostly used in tailored jackets.

7. Double pointed dart–Having two darts joined together at their widest ends, with one point toward the bust and other one toward the waist.

Double pointed darts

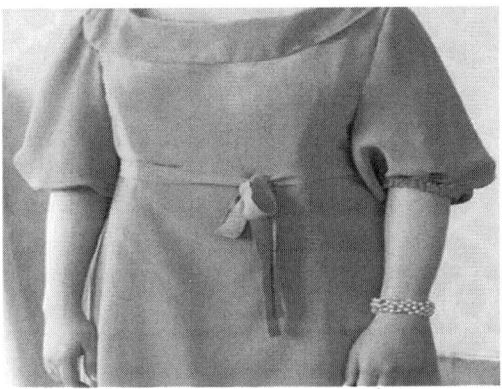

8. Curved darts–Curve darts used when the straight-line of a dart does not fit your shape. It can be slightly concave or convex depending on the need to fit.

1.3.1.2   Pleats

Pleats are introduced to provide fullness evenly all rounds. Each pleat provides different look based on how it formed. A pleat is an unstitched, folded dart held securely along the joining seam line. It is a fold in the fabric that releases fullness. Pleats are used as a design. Pleats are found on skirts, bodices, sleeves, dresses, jackets, etc. They are formed in a variety of ways. They may be folded and left un-pressed or pressed, stitched or left unstitched.

They may be grouped together with even or uneven spacing. Pleat depth may be single, doubled or tripled. Types of pleats are:

Knife pleats: Pleats are grouped and face in one direction.

Box pleats: Pleats are folded away from each other on right side of the garment.

Inverted pleats: Pleats are folded to meet each other on the right side of the garment.

Accordion pleats: Pleats have folds resembling the bellows of an accordion. The pleats are close together and depth is equal from waist to hemline.

Sunburst pleats: Pleats fan out and graduate from the waist. They are generally used on circular skirts.

1.3.1.3   Gathers

Gathers change the look of the basic garment, but do not affect the fit. Gathering is an effective and decorative way of distributing fullness over a given area. In dress making, gathers are often used at yoke lines, waist lines, and neck lines, upper and lower edge of sleeves. Types of gathers are:

- Gathers at shoulder
- Gathers at centre front
- Gathers at waist and
- Gathers at neckline.

Gathers at shoulder

Gathers at centre front

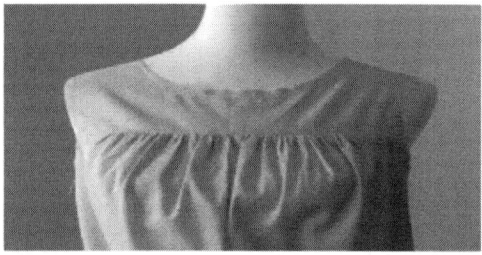

Gathers at waist

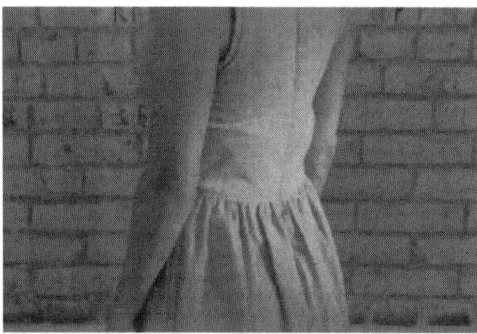

Gathers at neckline

1.3.1.4   Tucks

A tuck is stitched fold of fabric that is most often decorative, but it can also be a shaping device. Each tuck is formed from two stitching lines that are matched and stitched the fold of the tuck is produced when the lines come together. A tuck's width is the distance from the fold to the matched lines. Tucks that meet are blind tucks those with space between them are spaced tucks. A very narrow tuck is a pin tuck. Most tucks are stitched on the straight grain, parallel to the fold and are uniform in width. Curved dart tucks are an exception.

Plain tuck
These are tiny tucks used on baby clothes and fine blouses. They are usually less than
1/8″ wide.

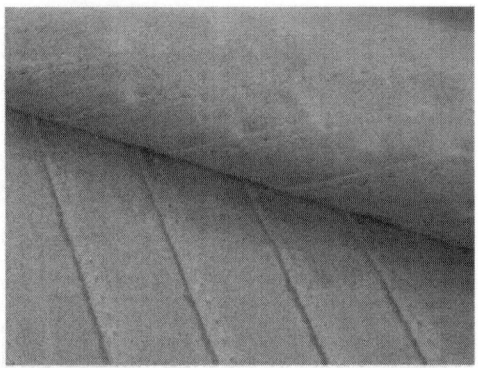

Corded tucks
These are made by placing cording on the wrong side of the fabric at centre of tuck before stitching the tuck. Stitching should be done close to cording.

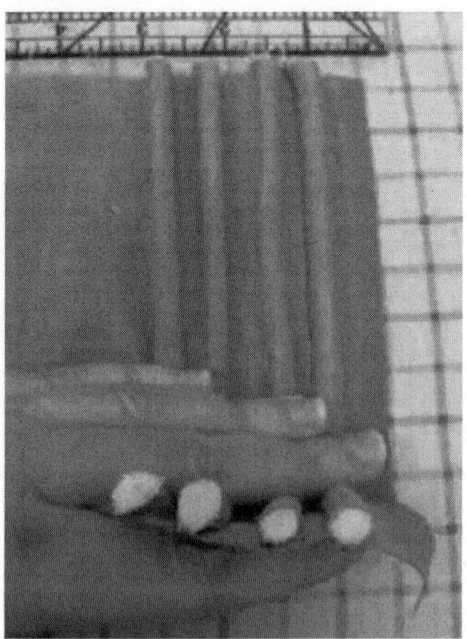

Shell tucks
This is very decorative tuck made by hand.

Cross-tucking
When rows of tucks are stitched along the fabric in both horizontal and vertical directions, the decoration is called cross-tucking.

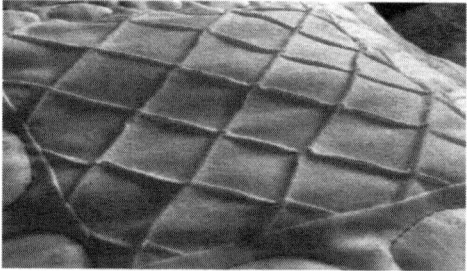

Group tucking with scalloped effect
Mark narrow tucks fairly close to each other, stitch them and press to one side. At right angles to the tucks, mark lines at regular intervals for crosswise stitching. On the first line stitch across the tucks in the direction they were pressed. On the second line stitch across the tucks after pressing them in the opposite direction. Continue stitching, alternating the direction of the tucks in successive rows.

1.3.1.5   Seams

Seams are result of joining together two or more pieces of fabric by means of stitching or fusing, but the basic function of a seam is to hold pieces of fabric together. To perform its function correctly, the seam should have properties or characteristics closely allied to those of the fabric being sewn. A "seam allowance" is the distance from the fabric edge to the stitching line from the edge. Seam allowance is planned according to the width needed for the type of seam, seam finish or garment design. There are only a few fundamental seams but by using a wide variety of finishes it is possible to adapt seams to materials of different weight and texture, to the different location and design of the garments so that the type of seam selected depends on the type of fabric, texture of fabric, the use and care of garment, placement or position of seam on garment. Most plain seams require a seam finish to prevent ravelling. A seam finish is a way of treating or enclosing the raw edges of seam allowance so they are more durable and do not ravel. Variations of the plain seam include bound encased, top stitched and eased seams. Some, such as the flat fell seam, add strength or shape. Others such as French or bound seams, improve the appearance of the garment or make it longer wearing.

Plain seam

A plain seam is the most basic and easiest to use. Its seam allowances are usually pressed open, although on lightweight fabric they can be trimmed and neatened together. In a well-made plain seam, the stitching is exactly the same distance from seam edge till the entire length of the seam. To ensure absolutely straight seam, it is advisable to practice stitching while keeping the fabric edge aligned with seam guideline on the throat plate of needle, it is basically used on fabrics that will not ravel like fine to medium weight cottons, linens or fine wools. On seams of garments that will be covered by a lining.

A straight seam is the one that occurs most often in most cases, a plain straight stitch is used for stretchy fabrics, and however a tiny zigzag or special machine stretch stitch may be used. It is rarely used for transparent fabrics such as voile, georgette, organdie, etc. It is frequently chosen for side seams in blouses, kameez and frocks, etc.

Steps of construction

1. Lay two layers of material together, right side facing right side.
2. Machine stitch at edge leaving an allowance of 1″. Start with back stitch and end with back stitch.
3. Press opens the seam, to avoid bulkiness and to make it smooth and flat.

Plain Seam

Curved seam

A curved seam requires careful guiding as it passes under the needles so that the entire seam line will be the same even distance from the edge. The separate seam guide will help greatly. To get better control, use a shorter stitch length (15 per stitch) and slower machine speed.

Steps of construction

1. Stitch a line of reinforcement stitching just on seam line of the curve.
2. Clip into seam allowance all the way to the stitching line at intervals along the curve.
3. Cut out wedge-shaped notches in the seam allowance of outer curve by making small folds in seam allowance and cutting at slight angle. Be careful not to cut into stitching line.
4. Press seam open over the curve, using tip of iron only. Do not press into body of the garment. If not press to contour, seam lines become distorted and look pulled out of the shape.

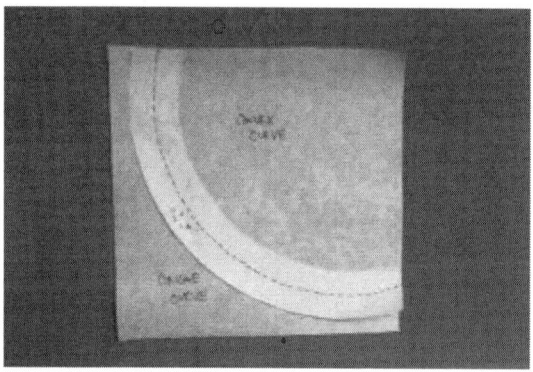

Corner seam

A cornered seam needs reinforcement at the angle to strengthen it. This is done by using small stitches (15–20 per inch) for 1″ on either side of the corner. It is important to pivot with accuracy when cornered seams are enclosed, as in a collar, the corners should be blunted so that better point results when collar is turned.

Steps of construction

1. To join an inward corner with an outward corner or straight edge, first reinforce the inward angle stitching just inside the seam line 1″ on either side of corner.
2. Insert a pin diagonally across the point where stitching forms the angle clip exactly to this point, being careful not to cut past the stitches.

3. Spread the clipped section to fit the other edge; pin in position then with clipped side up, stitch on the seam line pivoting at the corner.

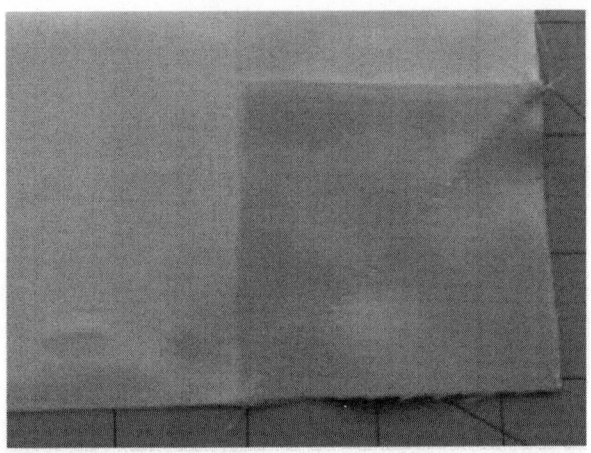

Seam finishes

A seam finish is any technique used to make a seam edge look neater and or prevent it from ravelling out. Though not essential to completion of the garment, it can add measurably to its life. Less tangibly, finished seams add a trim professional touch, in which you can take pardonable pride.

Plain straight seams are finished after they have been pressed open. Plain, curved or cornered seams are seams finished right after stitching, next clipped or notched, then pressed open. In this category we have the following seams: (i) Stitched and pinked seam (ii) Turned and stitched seam and (iii) Hong-Kong seam.

(i) Stitched and pinked seam

A seam finish in which a line of machine stitching is made ¼″ from the raw cut edge before pinking. It is done to prevent the pinked edge from ravelling, to prevent the seam from curling on fabrics which ravel slightly. It is a quick and easy finish suitable for firmly woven fabrics.

Steps of construction

1. Take two layers of fabric, right side facing right side; stitch on wrong side, leaving a distance of 1″ from edge. Press open the seam allowance (Straight plain seam).

2. Using a short stitch place a line of a stitching ¼″ away from the edge of the seam allowance. On the one side of seam allowance. Repeat the same on the other end of seam allowance.

3. Then pink the outer edge of the seam allowance away from the seam you have just applied.

4. Press opens the seam.

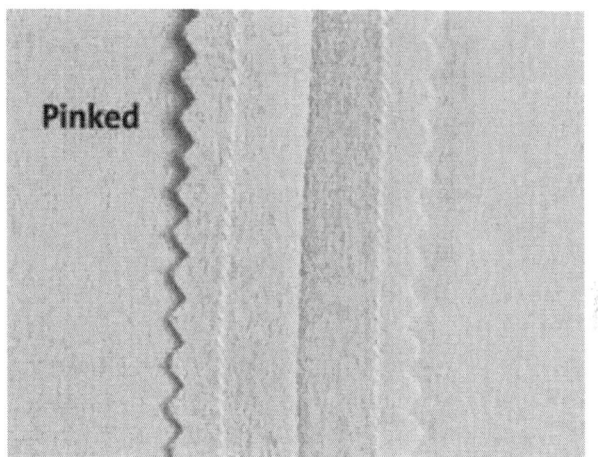

(ii) Turned and stitched seam

A seam finish in which the raw edge of the seam allowance is turned under stitched and concealed. Tailored edge, turned and stitched or clear finish all are the names of one seam. It may be helpful on difficult fabrics. This is a neat tailored finish for light to medium weight fabrics of cotton, linen and viscose. It is done to

• Prevent the seam edge from fraying

• On straightedge seams

• On garments where seam allowance will not show on the face of the garment

• On plain weave fabrics

• On unlined coat, jacket or vest.

Steps of construction

1. Take two layers of fabric, right side facing right side, stitch from wrong side at a distance of 1″ from the edge. Press open the allowance (Straight plain seam).

2. Turn under the edge of the seam allowance ¼″ stitch along the edge of the fold. Repeat the same step on the other edge of seam allowance.

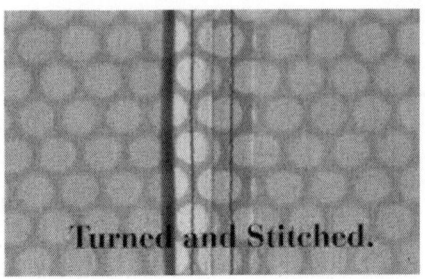

(iii) Hong-Kong seam

A seam finish in which the raw edge of the seam allowance is covered with a folded ribbon tape or bias binding. Hong-Kong seam is basically a couture finish on the hem edge, the Hong-Kong finish takes a little extra time and requires superior workmanship. This is done on heavy fabrics that ravel easily on the seams of unlined coats, jackets and vests. When the inside or wrong side of clothing may show,

- To reduce the abrasion of seam edge,
- To cover the raw edge of fabric that may safe the skin,
- To protect raw edge of easily frayed fabrics and on fabrics that are too thick to be turned under and edge stitched.

Steps of construction

It is also taken as an alternative to bias bound finish.

1. Right-side facing right side. Stitch at a distance of 1″ from the edge on wrong side. Press open the allowance.

2. Cut 1″ or 1½″ wide bias strip from a light weight material. With right sides together stitch bias strip to seam allowance ¼″ from edge.

3. Turn bias over edge to the underside and press. From the right side. Stitch in the crevice of the first stitching (Stitch in ditch) trim unfinished edge of bias.

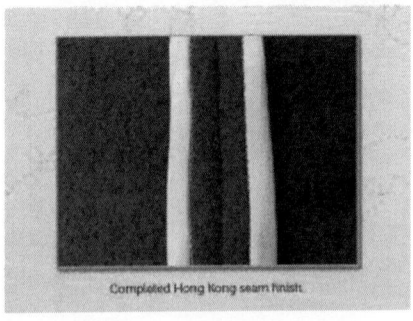

Completed Hong Kong seam finish.

Self-enclosed seams

Self-enclosed seams are those in which all seam allowances are contained within the finished seam, thus avoiding the necessity of a separate seam finish. They are especially appropriate for visible seams, such as occur with sheer fabrics and in unlined jackets. Also they are ideally suited to garments that will receive rugged wear or much laundering. Proper trimming and pressing are important steps if the resulting seams are to be sharp and flat rather than lumpy and uneven. Precise stitching is essential, too. This selection includes following seams: (i) The French seam (ii) Flat felled seam and (iii) Mock French seam.

(i) The French seam

A seam constructed so that a narrow seam is contained within a cage on producing a clear finish. This is a very secure and neat seam as the raw edges are not exposed. Since the finished seam consists of four layers of cloth, it is likely to be bulky. Hence it is suitable for thin/sheer fabric such as voile, organdie, and georgette. It is also used for dainty garments and lingerie. This is done to prevent fabrics from fraying where the seam finish will show through garments made of sheer fabrics (e.g., chiffon, organza, georgette and organdie). On children's and infants wear, underwear and outerwear. On straight seams when a seam is to appear a plain seam on the face of the garment and a clear finish is desired on the inside. It is not used in couture industry but is suitable for garments that require frequent washing e.g., night wear.

Steps of construction

1. Lay two layers of material together, wrong side facing wrong side. The first stitch is 1/8" or ¼" outside the fitting line, depending on the desired finished width of the seam.
2. Trim the edge so that it is less than desired finished width of the seam. It looks best when finished width is ¼" or less.
3. Press the seam in one direction. Turn the fabric so that right side is facing right side. Fold on the line of stitching. Machine stitch on the seam line. Since the raw edges are enclosed, this seam requires no special finish.

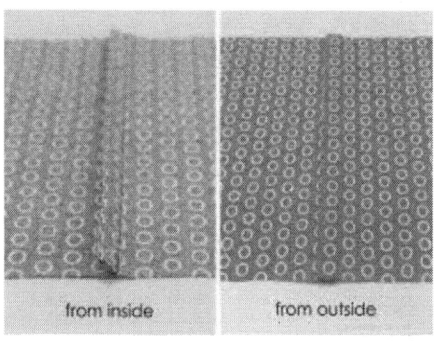

from inside   from outside

(ii) Flat fell seam

Place two layers of fabric with wrong side facing wrong side stitch from right side leaving an allowance of 1″ press open the seam. Trim inner seam allowance to ¼″. Press under the edge of the outer seam allowance which is trimmed to ½″. After pressing or folding outer seam allowance on inner one stitch this folded edge to the garment.

A flat felled seam is the result of enclosing both seam allowance by machining opposing folded edges beneath a row of machine stitches through all piles. The flat felled seam is very sturdy and so often used for garment that are made to take hard ware e.g., sports clothing and children's wear. Since it is formed on the right side, it is also decorative and care must be taken to keep the widths uniform within a seam and from one seam to another. Be careful to press like seams in the same direction (e.g., both shoulder seams to the front). Other examples are men's shirts, boy's trousers and women's tailored garment and unlined garments. Flat felled seams may be produced in all in on operation with a felling foot attachment on an industrial machine. In non-industrial production, seam may be made in two or more steps.

Steps of construction

1. Place two layers of fabric with the wrong side facing wrong side. Stitch from right side leaving an allowance of 1″. Press open the seam (Straight plain seam).
2. Trim the inner seam allowance to ¼″. Press under the edge of the outer seam allowance which is trimmed to ½″.
3. After pressing or folding outer seam allowance on inner one, stitch this folded edge to the garment.

(iii) Mock French seam

A plain seam made to resemble a French seam by the face-to-face enclosing of the folded seam edges. The mock French seam which is also known as

False French or Imitation French seam can be used in place of the French seam, especially on curves of armholes and princess line garments, where a French seam is difficult to execute on transparent fabrics that ravel easily and where a strong finish is required. Basically used for fabrics where two turnings are difficult to make, as in matching plaids.

Steps of construction

1. Take two layers of fabric, right side facing right side, stitch at a distance of ½″ from the edge on wrong side.
2. Turn in the seam edges ¼″ and press, matching folds along the edge. Stitch these folded edges together. Press seam to one side.

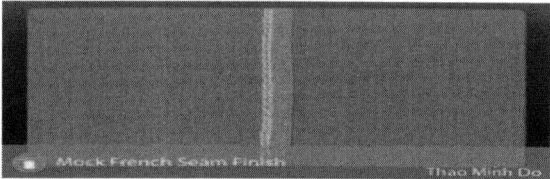

Topstitching seams

Seams are topstitched from the right side with usually one or more seam allowances caught into the stitching. Topstitching is an excellent way to emphasise a construction detail, to hold seam allowances flat or to add interest to plain fabric. There are two main considerations when top stitching. The first is that normal stitching guides will not, as a rule, be visible, so new ones had to be established. A row of hand basting or tape applied just next to the topstitching line can help. The pressure foot is also a handy gauge. The other consideration with topstitching is how to keep the under layers flat and secure even basting will hold pressed open seam allowances. Diagonal basting will hold those that are enclosed or pressed to one side. Grading and reducing seam bulk will contribute to smooth topside. A long stitch is best when topstitching used button hole twist or single or double strands of regular thread. Adjust needle and tension accordingly.

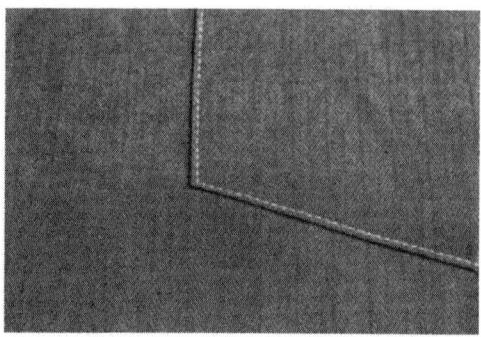

1.3.1.6   Necklines

Neckline is part of bodice, around the neck. In ladies garments, neckline can be shaped in different ways and styles to get a decorative effect. Necklines are perhaps the most conspicuous part of any dress design and for that reason deserve careful consideration, both while choosing a design as well as while sewing. Select a suitable and comfortable neck style and one suited to the fabric. The neckline can be of various shape and sizes. Some of the types of necklines are:

1. Boat neckline
2. Draw string neckline
3. Cowl neckline
4. Key whole neckline
5. Halter neckline
6. Scoop neckline
7. Square neckline
8. Heart shaped neckline
9. "V"-shaped neckline
10. Round neckline
11. "U"-shaped neckline
12. Wedge shaped neckline.

1. Boat neckline
It is a boat shaped neckline, approximately following curve of the collarbone, high in front and back, wide at sides, and ending in or at shoulder seams. Its front neck depth is generally kept more than the other neck depths.

2. Draw string neckline
It is a neckline with cord, threaded through a ceasing (i.e., folded edge with gap to insert tape) to be gathered and adjusted high or low. These are mostly used in kid's wear like, Jabla, Nighty and in ladies skirt top.

3. Cowl neckline
A cowl neckline is developed by adding one or more folds to the neckline cowls are always cut on the bias and have free and folded effect when worn cloth suitable for cowls are sheer and light weight type fabrics such as chiffon, silk, crepe, crepe silk, soft satin, georgette, jersey, etc. The pattern prepared for particular fabric, says satin, and cannot be used for chiffon, as each fabric is having different characteristics.

4. Keyhole neckline
It is a high round neckline with inverted wedge shaped opening at front. These necklines look good when it is finished with fitted facing. The

depth of round neckline and inverted wedge can be changed according to desire.

5. Halter neckline

It consist of a strap, rope, or band around neck, attached to backless bodice (i.e., back and shoulders bare) it is tied in a bow at back neck. It is generally used in kid's party wear and frocks.

6. Scoop neckline

It is a low curved neckline, it is cut deep in front or back or both. The shape of the neckline more or less resembles the shape of pot and hence is also called as pot neck.

7. Square neckline

This neckline shape resembles the square shape and has four corners, two in front and two at the back. This is used for kammez, frock, tops and other ladies garment.

8. Heart shaped neckline

It is a deeply cut neckline with its front lower edge in heart shaped curve. It is the variation of square neckline.

9. "V"-shaped neckline

It is shaped in front to a sharp point like the letter "V". On the centre front of the pattern, mark a point for the depth desired and draw a line to this point from the neck end of the shoulder line.

10. Round neckline

Because a round neckline is wider in effect than a "V"-shaped neckline, it is usually cut somewhat higher, its depth on the pattern being determined as suggested for the "V"-neckline when the depth has been noted, draw a line from the shoulder to this point, having the line extend downward with a slight slant for about two-thirds of the neckline depth and then curve abruptly towards the centre front.

11. "U"-shaped neckline

It is cut in front in the shape of letter "U". It is the modification of round neckline. The depth of neckline is more than the normal round neckline.

12. Wedge-shaped neckline

This is another variation of neckline where a straight line and curved line is combined to form a wedge-shaped opening. Mark points to indicate the depth wanted and the width at lower line, which is wider than the regular square opening, join these points by a straight line, and then draw the side line in the curved effect joining the straight line.

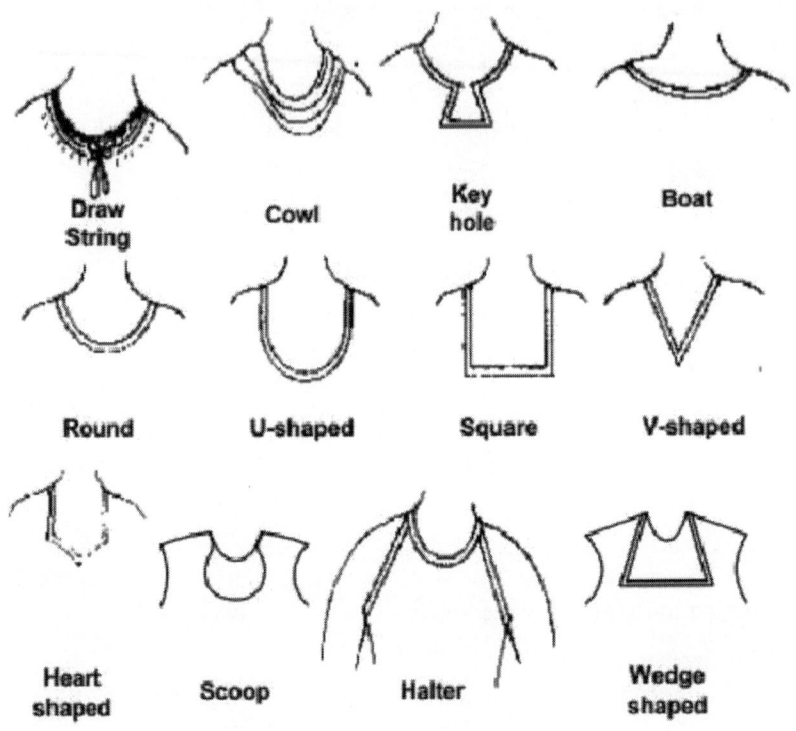

Draw String · Cowl · Key hole · Boat

Round · U-shaped · Square · V-shaped

Heart shaped · Scoop · Halter · Wedge shaped

Neckline used in men's T-shirt

- **Crew neck**–The most popular style for men's T-shirt, the round neck or crew neck is the most traditional and common.
- **Deep scoop**–The scoop neck had a deep cut and is mainly used by guys, who want to show their muscles.
- **Deep V-neck**–The deep V-neck ultimate aim is showing-off your body.
- **Scoop neck**–Similar to crew neck collar are more oval. Past 3–4 years this neck line gained more popularity.

Neckline finishing

The neatness, with which the neckline is finished, is of vital importance, for it greatly affects the final appearance of the garment. Besides the neckline under goes much strain during putting on or taking off the dress. Hence the mode of finish must be such that it allows one to retain the desired shape, pattern and must be durable. Bias strips are generally used to finish the neckline, because of its stretch ability. Necklines are usually curved and hence tend to stretch during handling. A stretched neckline can spoil the appearance of garment. So great care must be taken in handling and finishing neckline.

Most necklines are finished in one of three ways—with facings, bindings or with a collar. Regardless of the type a well-made neck finish should conform to these standards: Neck edges must not be stretched out of shape.

All seams and edges must be as thin and smooth as they can be made, without weakening the garment. All edges must have good shape with accurate corners and curves. Facing edges, whether beneath a neckline or collar, must be hidden from view. Body and stiffness of collar must be sufficient for the design.

Bias piece

True bias which is otherwise called as cross piece or falls on a diagonal line at 45°C to the lengthwise and crosswise grains. It has the maximum elasticity or in other word it stretches more than any other direction on cloth. True bias is used to finish raw edges. It is useful especially for finishing curved edges such as neckline, sleeveless armholes and scallops. A straight piece of material attached to a curve will look bulky and untidy. The elasticity of bias permits it to stretch or contract and thus takes the shape of any curved edge giving it a flat smooth finish. Bias strips can be applied as facings and bindings.

1. Cutting bias strips

Fold the fabric diagonally so that the lengthwise threads of the folded part fall parallel to the crosswise threads on the rest of the material. (If the grain lines cannot be clearly seen, mark the lines with chalk first.) Using a gauge or ruler, measure from the fold to desired width of bias strip (usually 1 to 1″) and draw parallel lines. Cut strips along the marked lines and trim-off ends along the warp threads.

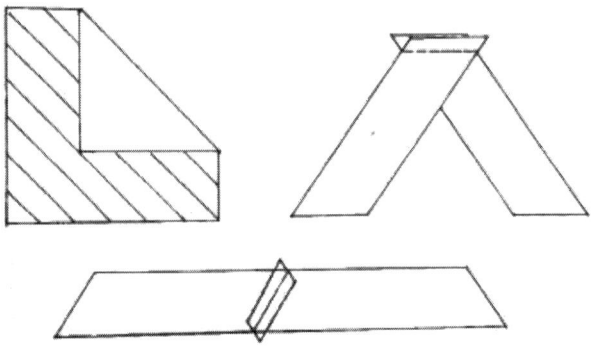

Marking, Cutting & Joining of Bias Piece

2. Joining of bias strips

Place the two strips to be joined right sides facing and the edges of the cut ends coinciding. The strips will now be at right angles to each other. Shift the top strip beyond the other so that the sharp points at the ends of the strips project on either side. Stitch a seam joining the points where the sides of the two strips intersect. Press the seam open and trim the seam projection showing on right side.

Facing

These are used to provide a neat finish to the raw edges in a garment and to support the shape of neck line, armholes, collars, etc. When the edge to be faced is a straight line, the facing may be cut in one piece with the garment section. If the shape of the neckline is a curved one a bias piece can be used. Usually facing is applied separately. The colour of the facing piece must co-ordinate with the colour of the garment fabric.

Facing appear on the right side of the garment. The right side of the facing must be matched to the wrong side of the garment to ensure that it will be right side out when finished. If this is to be applied to the neck line, shoulder seam of the garment, it should be reverse just inside the outer finished edge of the facing. This is to prevent raw edges of shoulder seam from showing at the neck line. Decorative facing are usually made with scallops, points or other designs along the outer edge. Particular care should be taken to see that the right and left halves are symmetrical in design and shape.

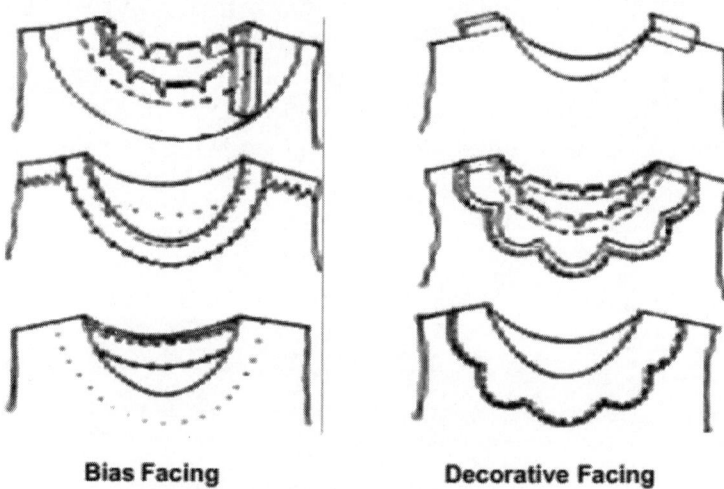

Bias Facing   **Decorative Facing**

Bias facing can also be applied on the right side of the garment for decorative effects, but no edge designs are possible with this facing because it cannot be made wide enough. The decorative effect has to be obtained by the use of material in some harmonising colour with prints like checks, stripes, dots, etc.

Binding

Bias binding is used to finish and strengthen raw edges and to add a decorative trim to a garment. It shows both on the right and wrong side. It is used to finish necklines, armholes, sleeve edges, front closings, collars, cuffs and seams. It can be adapted equally well to straight, curved gathered and irregular edges (like scallops). When finished, bias binding should have uniform width and should lie flat and smooth without any stitches showing on the right side of the garment. There are two kinds of bias binding. Single binding and French binding or piping. Binding may be prepared or may be bought as commercial bias binding.

1. Single bias binding
Cut a bias strip that is twice the finished width plus two seam allowances. Tack the strip to the garment right sides facing. Remember that bindings are handled in the opposite manner to facings at inward and outward curves. For bindings, stretch the bias on inward curves and ease it on outward curves. Stitch the binding to the garment with a plain seam. Trim the seam as wide as the finished binding. Turn under 1/8"–1" on the outer edge of the bias and fold it over the seam on the wrong side. Now hem the fold to the line of stitching using hemming stitches.

2. French binding or piping
French binding or piping is used on sheer fabrics. For this you must cut bias strip that is six times the desired finished width. Fold the strip in half, wrong sides together and press. Stitch raw edges of binding to the garment on the right side and hem the folded edge to stitching line on the wrong side. It is also called double binding.

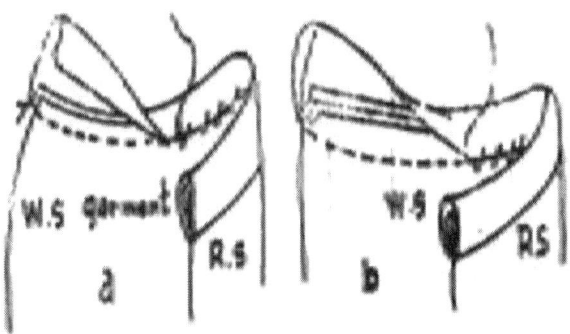

Single & Double Bias Binding

1.3.1.7   Collars
A collar is the part of a garment that encircles the neck and frames the face, offering great opportunities for design variations. Collars can be developed close to or away from the neckline. They can be wide, narrow, flat, or high, and with or without an attached stand. The collar edge may be stylized or may follow a basic shape (round, curved, scalloped, square, pointed, etc.).

Collars can be convertible (can be worn closed and open, so that it lies flat across the chest when opened) or non-convertible (stay in the same location whether garment is buttoned or unbuttoned).

Different types of collars

1. Mandarin—A small standing collar, open at the front, based on traditional Manchu or Mongol-influenced Asian garments.
2. Band—A collar with a small standing band, usually buttoned, in the style worn with detachable collars.
3. Stand—A standing collar is a short unfolded stand-up collar style on a shirt or jacket.
4. Tuxedo—The winged collar is almost entirely reserved for formal occasion dress. Wearing a winged collar requires it to be paired with a tuxedo or a cutaway suit. A bow-tie fits the outfit and the occasion.
5. Pilgrim—The definition of a pilgrim is a person who travels somewhere for religious reasons.
6. Chelsea—A Chelsea collar is a "V"-shaped neck with a straight collar set into it, the collar meeting at the front.
7. Peter Pan collar—A Peter *Pan collar* is a style of clothing *collar*, flat in design with rounded corners.
8. Shawl—A Shawl collar for a V-neckline that is extended to form lapels, often used on cardigan sweaters, dinner jackets and women's blouses.
9. Bertha—A Bertha collar is a wide, round, flat collar designed to accent a woman's shoulders. It has a long history stretching back to Victorian fashion. It can be worn as an accessory to a dress or a top.
10. Sailor—A collar with a deep V-neck in front, no stand, and a square back, based on traditional sailor's uniforms.

11. Fichu—A collar styled like a fichu, a large neckerchief folded into a triangular shape and worn with the point in the back and the front corners tied over the breast.
12. Notched—A wing-shaped collar with a triangular notch in it. Often seen in blazers and blouses with business suits. Also, rounded notched collars appear in many forms of pyjamas.
13. Wing—A small standing collar with the points pressed to stick out horizontally, resembling "wings", worn with men's evening dress.
14. Polo—A high close-fitting turnover collar.
15. Henley—An Henley shirt is a collarless pullover shirt, characterised by a placket beneath the round neckline, about 3–5″ (8–13 cm) long and usually having 2–5 buttons. It essentially resembles a collarless polo shirt.

1.3.1.8   Sleeves

Sleeve is that part of the garment, which covers the arm of the body and is usually attached to armhole of bodice pattern. Sleeves support the design and functional element of a garment. They are broadly classified into three types–set-in sleeves, raglan sleeves and kimono sleeves which are further made into separate styles. Sleeve is that part of the garment, which covers the arm of the body and is usually attached to armhole of bodice pattern. Sleeves support the design and functional element of a garment. In design sleeves should complement the bodice of the garment and as for functional sleeves should provide ease of movement and comfort. In today's world not only does the garment vary in designs and styles but the sleeves too have different styles and thus vary in their construction. By choosing a sleeve style that suits the figure of wearer, design of the fabric, design of the dress and current fashions, it can enhance the appearance of the dress.

1. **Puff sleeve**—Puffed or puff sleeve, a short, ¾ length or full sleeve that is gathered at the top and bottom, now most often seen on wedding and children's clothing.
2. **Leg of mutton sleeve**—The **leg** of **mutton sleeve** (also known in French as the gigot **sleeve**) was initially named due to its unusual shape: formed from a voluminous gathering of fabric at the upper arm that tapers to a tight fit from the elbow to the wrist.
3. **Petal sleeve**—A two piece **sleeve** that overlap to form a **petal** shape.
4. **Peasant sleeve**—A full sleeve gathered at top and bottom.
5. **Juliet sleeve**—A long, tight sleeve with a puff at the top, inspired by fashions of the Italian renaissance and named after Shakespeare's tragic heroine, popular from the empire period through the 1820's in fashion.

6. **Kimono sleeve**—Kimono sleeve are wide loose short or wrist length. This style is obtained from **traditional Japanese dresses** and adapted as today fashion.
7. **Bishop sleeve**—A long sleeve, fuller at the bottom than the top, and gathered into a cuff (1940's).
8. **Bell Sleeve**—A long sleeve that is fitted from the shoulder to elbow and gently flared from elbow onward.
9. **Raglan sleeve**—A sleeve that extends to the neckline.
10. **Dolman sleeve**—A long sleeve that is very wide at the top and narrow at the wrist.
11. **Lantern sleeve**—A plain at top and wrist ballooned out half-way between the wrist and the elbow. Cut in two pieces with a seam going around the sleeve at the fullest part.

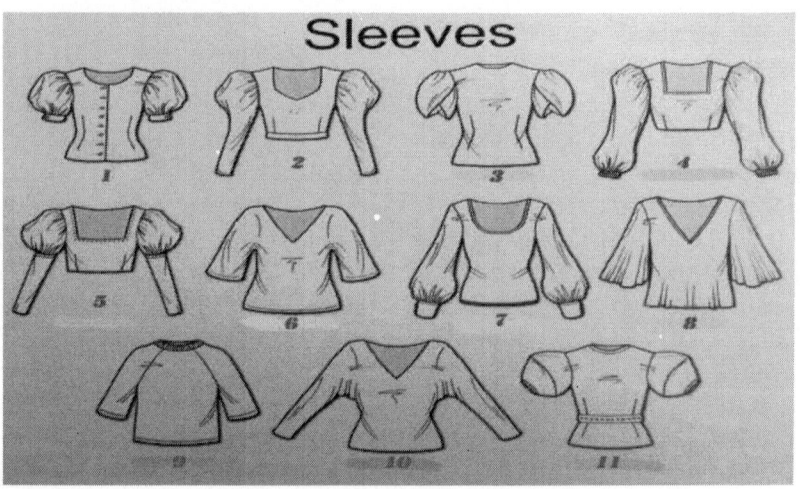

1.3.1.9   Plackets

Plackets are particularly seen in women's garments which have a good fit. They are usually found near the writs, neckline, waist line and other comfort fitting parts of the garment. These are finished openings which are normally kept closed (to have a tight fit) and are opened only while wearing or taking-off the garment. Fasteners like zippers, buttons, buttonholes, hooks are used in keeping the plackets closed. The term placket was first used in the 16th century to describe an underskirt. Choose opening was in turn, called the placket hole. Today the term describes any finished slit (or) in opening in a garment.

Plackets are finished opening constructed in order to make it easy to put on or take-off a garment. When the garment is in use plackets are kept closed (for good fit) with the aid of fasteners such as zips, buttons and buttonholes, press buttons, hooks and eyes, etc. They are used at waist lines, neck lines, wrists and other snug fitting parts of the garments.

A placket may be made in an opening left in a seam, in slash or cut in a garment. The placket made in a seam is stronger and gives a better finish when completed. The following points should be kept in mind while making placket. A good placket should be as inconspicuous and flat as possible. It should not be bulky, puckered or stretched. Fastenings should hold securely and there should be no gapping edges. Neck openings must admit the head easily, pass over. The position of the placket should be such that it is easily accessible and convenient to operate. Openings are subjected to certain amount of strain during wear and should be strengthened at the closed ends, lower end or neck and skirt openings, upper end of wrist openings, etc. For plackets in seams to be durable, the garment seam should be at least 5/8″ wide. Seams should not be trimmed or clipped too close. The type of placket used should be suitable to the kind of garment on which it is used, its position in the garment, texture of the fabric, age and sex of the wearer and current fashion must be kept in mind while choosing placket.

Type of Plackets
Front plackets

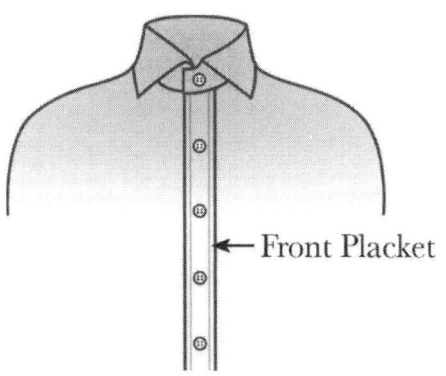

←Front Placket

In a front placket, the extra material of the placket is sewn in front of the shirt material. The buttons poke through from the back and rest on the front placket. Typically there is a stitching on both sides of the placket to provide additional support. This is by far the most popular type of placket in North America. You're most likely to see the front placket on more casual button-up shirts—the kind you might wear with a pair of slacks. Front plackets are quite popular on more formal dress shirts, too.

French placket

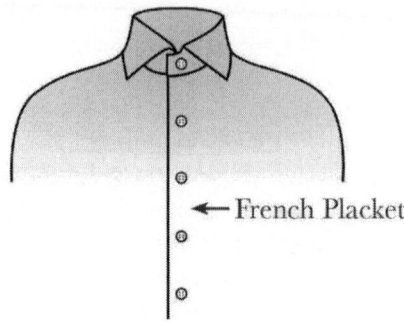

Instead of having the extra support in front of the shirt material it is sewn in behind. The additional support of the extra material is still there—you just don't see it with the French placket.

Concealed placket

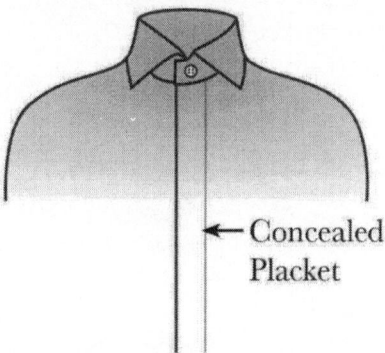

As the name indicates the placket is hidden and you don't actually see the buttons holding the two sides of the shirt together. There is a rear strip of reinforced material that the buttons loop through and then a front piece of material that cover the buttons. The concealed placket is the global standard for tuxedo shirts, but you'll also notice it on that black evening dress shirt you've been planning to buy so you look as good at the hottest club in town as you do in a board room power meeting.

1.3.1.10   Skirts

Types of skirts based on their lengths

Skirt is something that covers the lower part of the body especially for girls, women or ladies. Skirts come in a variety of shapes and styles and they are essential items in women's wardrobes. Skirts are more elegant and easy to wear and besides they can be worn in all occasions. It is easier to design skirt than any other garment. In today's fashion world, we find a variety of types of skirt which can be categorised based on size, fabric, and design. The silhouette can be narrow, flared or bouffant. Even straight skirts can vary from extremely tight and narrow at the hem to slightly A-lined. Here is a brief description of the variety of skirt types being used on their vast types.

- **Normal skirt**

It is a normal skirt which is 2–3″ below the knee size.

- **Micro-miniskirt**

These skirts have length which extends 10″ from the waist above the mid-thigh.

- **Miniskirt**

The skirts whose length ends at mid-thigh level, are called miniskirt. These are very short and are usually long enough to reach between the crotch and the knees i.e., about 3″–5″ below crotch. They are casual or semi-dressy. In dressy or formal settings, they can be inappropriate.

- **Ballerina skirt**

The skirt which ends at mid-calf level is called ballerina skirt.

- **Maxi skirt**

Maxi skirts are those skirts whose length extends till the ankle or floor. These lengths vary a little based on vary to fashion and customers taste.

- **Broomstick skirts**

Broomstick skirts are either knee-length or ankle length and they have a wrinkled appearance. They are characterised by the three, or more, horizontal seams that wrap around the skirt in layers. They are either casual or semi-casual but cannot be dressy.

Skirts based on design

Based on invariable designs the skirts can be classified into the following types:

- **Basic skirt**

The basic skirt will have the darts of equal lengths in the form of fullness at the waist line in the front and back. The pattern obtained can be used as the base for any other skirt with a hemline sweep that is equal on the front and back.

- **Pleated skirt**

A pleat is a fold in the fabric that releases fullness. These types have lots of pleats of similar size around the waistline. It helps in giving an extra fullness to the skirt. Pleated skirts have vertical pleats running up and down the skirt all the way around. When there is movement, the skirt appears to open up.

Pleats can be of various types and therefore have specific names for specific pleat based on their design for e.g., box pleated skirt, inverted pleat skirt, sunburst pleated skirt. Knife pleated skirts and, accordion pleated skirt. These either reach to the knee or above the knee, and they require ironing. Pleated skirts are casual or semi-dressy.

- **Skirt with least flare**

This type of skirt fits the waist tightly with only two darts and has only 1"–2" ease allowance on the hipline. About 2" flare is added on the side seam of skirt for freedom of movement.

- **Circular skirt**

As the name suggest, this skirt looks like a circle when open fully therefore a fabric having maximum width is required for these kinds of skirt. The fabric is cut in a circle, like a donut. Then the elastic is sewn on top to create a waistband. The circle hangs right at the waist and allows the fabric to drape beautifully around you. These skirts do not need any side seam.

- **A line skirt**

A line skirts represent the shape of the capital letter "A" hence the name i.e., an A line skirt is a skirt that is fitted at the hips and gradually widens towards the hem, giving the impression of the shape of a capital letter A. The waistline measurement remains the same as in plain skirt but fullness is added to hem so that its circumference increases. If more fullness is added at the waistline then it becomes a flared skirt. It is also one of the easiest garments for a beginner to make. It has only three pattern pieces a front, a back, and a waistband.

- **Gored skirt**

A gore is a triangular piece of fabric. A gored skirt is one with gores which are narrow at the waistline and wider at the hemline. It can have any number of gores which can be equally or unequally spaced as "desired" by the wearer. The gore can be of various types such as angled, flared, and pleated or may be straight from the hip level. There are two basic types of gored skirts, they are 6-gore and 4-gore skirt. These kind of skirts are preferred for bulky fabrics where in on finds difficult to gather the fabric at waist.

- **Gathered skirt**

Gathered skirts are as the name suggests skirts that have gathers at the waistline. Usually thin fabrics are used for making these types of skirts as they can be gathered easily and there is no difficulty in stitching at the waistline where as bulky fabrics can have limited fullness as there will be difficulty in stitching it to the waistline. For a good gathered skirt the length of fabric according to the length of the skirt +2" (for folding) and width of the skirt must be twice the amount of waist circumference for gathering.

- **High- and low-waist skirt**

Low- and high-waisted refer to where the garment's waistline is meant to sit relative to one's torso. For example, low-waisted skirts are designed to sit on the hips where as high-waisted skirts are designed to sit higher than the belly button. Low-waisted skirt is 3½″ down from the natural waistline and a high-waisted skirt has extended waistline at any desired amount. The low-waisted hugs the hipline below the waistline whereas the high-waisted skirt extends above the natural waistline.

- **Layered skirt/Tiered skirt**

Also called as tiered skirt layered skirts are skirts which have layers of fabric attached to the each other at the hemline. The length and width of each layer may be same or may vary. Each gives a different look. The fabric used may or may not be the same. Frills can also be attached.

- **Pencil skirt**

Pencil skirt lengthens from the waist to beneath the knees or down to mid-calf. It hugs the body and is usually made from stretchable fabric. These kinds of skirts are preferred by slim figured women or girls. They are straight cut therefore makes it difficult to walk in. These are dressy or formal skirts. The hemlines are decorated by adding coloured ribbons, fringes, etc.

- **Pegged skirt**

It is also known as peg-top or inverted skirt as it has fullness introduced at the waistline and with no fullness at the hemline. The skirt may be gathered, pleated into the waistband.

- **Bubble skirt**

Create a bubble effect at the bottom of the hem which is tucked back under. The bubble skirt consists of an outer skirt which is gathered onto a lining. Bubble garments look best in rather light fabrics such as (double) gauze and lawn.

- **Skirt with peplum**

Peplum skirts are an extra overskirt flounce sewn onto, and dropping from the waistline or waistband. Peplums extend from the waist, down till the hip.

- **Godet skirt**

A godet is an extra piece of fabric in the shape of a circular sector which is set into a garment, usually a dress or skirt. The addition of a godet causes the article of clothing in question to flare, thus adding width and volume. Adding a godet to a piece of clothing also gives the wearer a wider range of motion.

1.3.1.11   Yokes

Introduction

A **yoke** is a shaped pattern piece which forms part of a garment, usually fitting around the neck and shoulders, or around the hips to provide support for looser parts of the garment, such as a gathered skirt or the body of a shirt. The yoke of a dress shirt is the area under the collar that drapes over the shoulder and holds the shirt's backing over the body. It is essentially the piece of the garment that behaves as a hanger, and creates the crisp lines of the shirt's backside. If a yoke is not well-made, it can cause gathering of the fabric in the centre of the back, which makes the back look pinched, and often cheapens the entire look of the dress shirt. They are usually sewn with a double layer, so the shape has correct fabric weight to drape over the body, and does not allow any wrinkling.

Definition

A yoke is a segment of a garment usually placed at the shoulders, above the waistline-at midriff or below the waistline-at hip. There are two basic types of yokes.

- Midriff yoke
- Partial yoke.

Midriff yoke

Referred to as torso or waist yoke and is a good device for securing fullness over the bust and provides a smooth and trim fitting around the waistline.

Partial yoke

A yoke, which does not extend across the entire garment, is called a partial yoke. They are used in:

- For controlling and supporting fullness needed over the bust, chest, hips, etc.
- To keep the upper area or the waistline of the garment trim and smooth.
- For decoration and may not have any fullness.

The depth of a yoke is usually established by reference to the CF line and not to the armhole, and it is marked as a proportion of the CF length. "Balance marks" have to be drawn across the yoke line to fix the position of the fullness under the yoke.

1.3.1.12   Pockets

Pockets are of different styles and shapes, some pockets are designed to be used for filling things inside, these types of pockets are referred to as functional pockets, others are made as decorative style, and some other pockets are hidden in view.

It doesn't matter what type of pocket that are on garments, what matters most is that some pockets on any apparel give the wearer either a professional or casual look.

Pockets can be designed using the same fabric as the garment or use a different colour to give it a contrast look. Some pockets are with flaps, while others are top-stitched with no flap.

These are some of most common pockets you can find on garments.

Patch pocket
A pocket that is pressed and sewn on to the exterior of a garment.

Patch with pleat
As the patch pocket, but with a box pleat to create more space within the pocket.

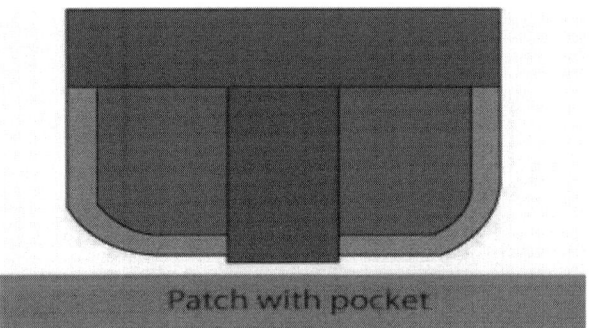

Jetted pocket
The pocket is constructed by cutting through the garment to the required length of the finished pocket, then the edges are bound and a pocket bag attached to the back of the garment.

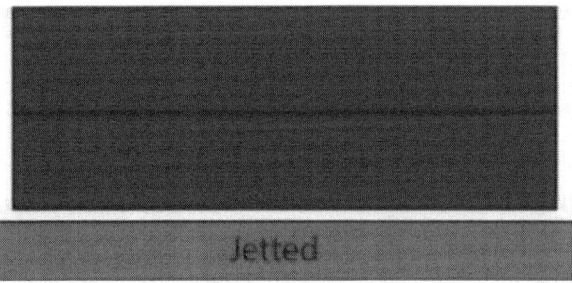

Bound patch

As the patch pocket, here shown gathered into a binding applied to the top edge to neaten it.

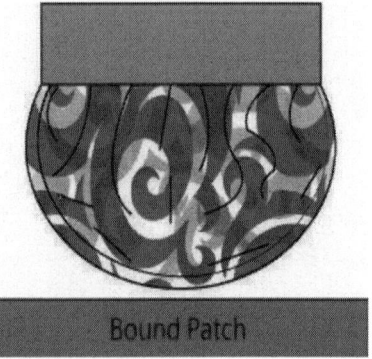

Shirt

Normally a breast pocket placed on any kind of shirt but usually a work shirt. It is a patch pocket with a shaped bottom and a turned back and top-stitched welt effect at the top.

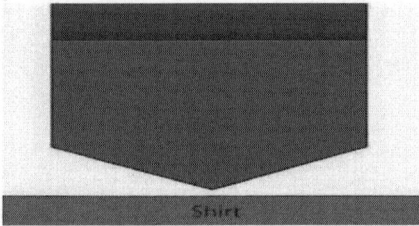

Patch with flap

As the patch pocket, but with a bagged out flap, the same width as the patch and stitched above the patch, to cover the opening. It is finished with a button or stud fastening.

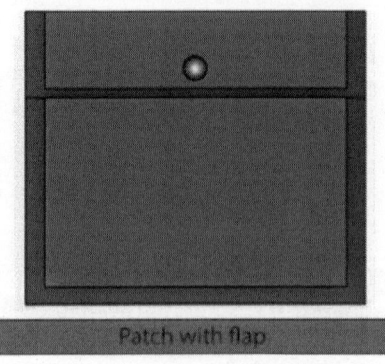

Welt pocket

In the garment is slashed to the length of the finished pocket and a folded and bagged out piece of fabric, the width of the finished pocket, plus seam allowance, is set into the slash and stitched up at the sides. The extended flap is stitched down at the sides and covers the pocket opening.

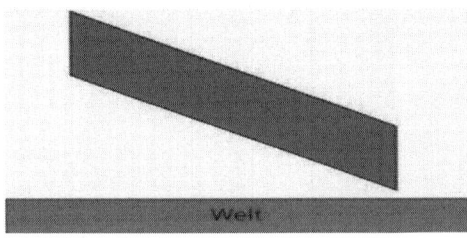

Welt

Shirred patch

As the patch pocket but the head of the pocket is elasticised to create a more spacious pocket.

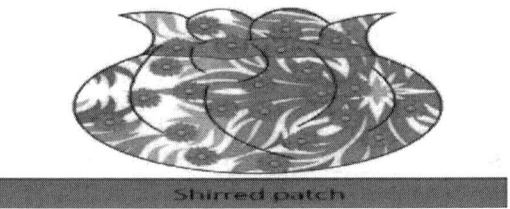

Shirred patch

Double pocket

This is a patch pocket that is layered to create two pockets. The zipped top is the entrance to one pocket and here the left side is the entry for the other.

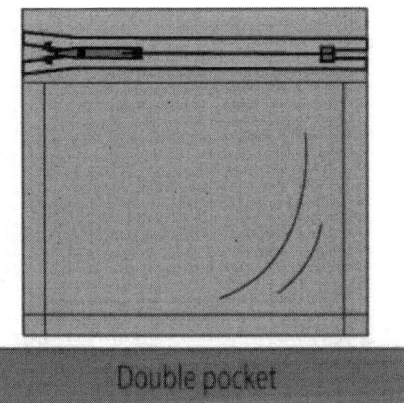

Double pocket

Western pocket
Like the angled flap, but with a bottom carving to a point, echoing the western or cowboy style of pocket.

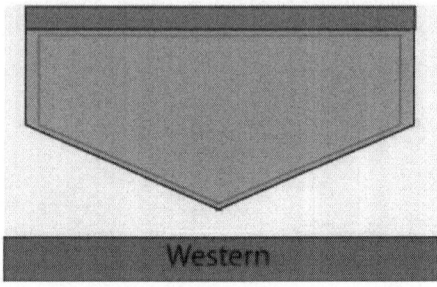

Rounded flap
Like the angled flap but with curved edges.

Petal pocket
A patch pocket that is split in two and overlapped with a curved top, to create a folded petal effect.

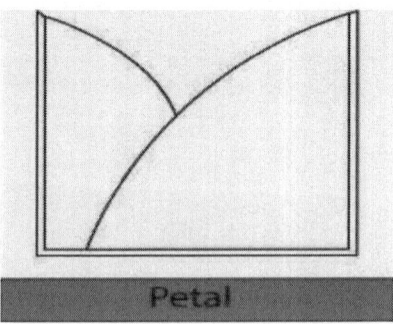

Side pocket
This pocket is set into the side seam of the garment, similar to the hidden in seam pocket.

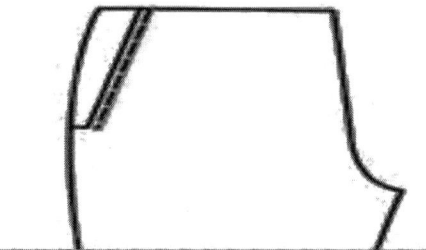

Cargo pocket

Similar in construction to the mechanic's pocket but applied to the waist of jeans or dungarees. The belt passes through the top of the pocket.

1.3.1.13   Fasteners

Button and decorative snap closures—Buttons and buttonholes are one of the most common methods used to join two pieces of a garment. In women's clothing, buttons are placed on the left side of the opening and the buttonholes are placed on the right overlap. In men's clothing, buttons are placed on the right side. The under lap and the overlap must be at least one half the button diameter or snap width plus one-fourth of an inch beyond the centre front or the closure seam line.

Buttons (including decorative snaps)

1. Buttons co-ordinate with the garment's design, fabric and garment care.
2. Buttons are spaced appropriately for their size and location.
3. The fabric under the buttons is additionally reinforced when necessary.
4. The buttons are sewn securely.
5. No loose threads hang from the buttons.
6. The buttons have a self or thread shank appropriate to the fabric's thickness.

Buttonholes

1. The type of buttonhole is suitable for the garment's design and fabric.
2. The buttons and the buttonholes are aligned so that the button rests within the top 118″ of vertical buttonholes, and within 118″ of the centre front of horizontal buttonholes.
3. The buttonholes are securely stitched in thread that matches or decoratively contrasts with the fabric. Hand or machine stitching is regular and smooth in appearance, with no fraying or loose ends.
4. The buttonholes are large enough to allow the buttons to pass through easily and yet small enough to hold the garment closed.
5. The buttonholes are even in length, width, and equally spaced unless otherwise designed.
6. If bound, the buttonhole must have the following criteria:
   a. The rectangle has perfectly square corners.
   b. The rectangle's length and width are determined by the button size and fabric weight.
   c. Welts are even in width and meet exactly at the centre of the opening.
   d. The facing is securely fastened to the back of the buttonhole.
   e. For pressing, see the description under the pressing section.

Snapped and hooked closures—Some varieties of snaps and hooks are used in concealed applications, while others are used in decorative as well as functional applications.

1. Fasteners are the correct size and location for the closure requirement. Sets are aligned and hooks are usually placed 118″ (3 mm) from the edge of the overlap so the garment is secure and the closure lies flat.
2. Fasteners are attached securely and neatly.
3. Concealed applications of fasteners are inconspicuous.
4. Fasteners used in visible applications are suitable for the garment design and fabric.
5. Durable coverings (thread or fabric) are used where appropriate.
6. The garment is reinforced on the wrong side, usually with interfacing.

Zippered closures

1. The zipper type and application are suitable for the garment's quality, design, fabric and use.
2. The zipper length is adequate for ease in wear or use.
3. Any visible stitching is straight, even and the thread matches, unless otherwise designed.
4. The zipper is securely inserted into the garment at the intended position.

5. The zipper, when closed, is flat and smooth, free from puckering and does not buckle.
6. The zipper opening appears to be a continuation of the garment's seam line.
7. Horizontal seams meet across the zipper opening.
8. Facings at the top of the zipper roll to the underside. Edges are smooth, even and flat.
9. The lapped zipper covers the stitching on the under lap so that the stitching is not visible.
10. The slot zipper is centred. Welts on each side of the placket appear identical in size, shape and placement, as well as equidistant from the opening.
11. The zipper slides easily and does not catch.
12. Fabric patterns are matched appropriately.
13. The seam at the end of an invisible zipper is smooth and straight.

Decorative detail
The trim enhances the garment or makes it unusual in some way, without overpowering the garment's design.

Soft trims—Soft trims include items such as lace, braid, ribbon, piping, and bias binding.
1. The trim is suitable to the garment fabric's weight, design, and care requirements.
2. The trim is securely attached to the garment.
3. The trim is attached in an inconspicuous manner, unless the method of attachment constitutes part of the decorative effect flexible trim is used on curved areas and applied without stretching or puckering of the trim or the garment.
4. Trims used at the comers are metered or appropriately applied to lie flat.
5. Bias binding and piping lie smooth with no rippling. See neckline treatments for more specific reference on bias binding.
6. There is no excess bulk at the joins or the ends.

Hard trims—Hard trims include decorative items such as buckles, belts, studs, beads, and sequins.
1. The hard trim is compatible with the garment fabric's weight, style, and care requirements and will not damage the garment.
2. The hard trim is securely attached.
3. Beads, sequins, and studs are applied so that the fabric does not pucker, and the underside application is smooth.

4. Belts meant to be firm have a stiff backing which is securely attached and does not show on the face of the belt.
5. Belts do not rub off or bleed colour onto the garments with which they are worn.
6. The belt buckle is securely attached to the belt and holds the free end of the belt securely when closed.

Fabric and stitchery trims—This category includes self-fabric and co-ordinating fabric trims such as ruffles and bows, appliqués, and decorative stitchery.

1. Ruffles are neatly finished and smooth and have ample fullness, even gathers, and no puckers or pleats.
2. Appliqués are securely attached to the base fabric, with no puckering, ravelling or fraying apparent.
3. Fabric bows are neatly turned, with no seam wells and with symmetrical ends.
4. Fabric flowers are neatly finished with no raw edges and are securely attached.
5. Topstitching is an equal distance from the edge at all points. An appropriate stitch length is used, and all loose thread ends have been hidden.
6. Decorative stitchery does not distort the garment.
7. The thread used in the stitchery is colourfast.
8. Care requirements of the appliqués and the stitchery are compatible with those of the garment.

1.3.1.14   Colours

Colour is simply light of different wavelengths and frequencies and light is just one form of energy made up from photons. Colour is a visual language. It is personal and universal sending a message of endless variation. It evokes our feeling and affects our life. The apparel of colour is to use it beautifully.

Colour theory
1. Prang colour chart
Basically colours are divided into three groups.

1. Primary colours
2. Secondary colours
3. Tertiary colours.

There are three primary colours: RED, YELLOW and BLUE. Primary colours cannot be created by mixing other colours. Secondary colours can only be created by mixing true primary colours. Tertiary colours are combinations of primary and secondary colours. There are six tertiary colours: red–orange, yellow–orange, yellow–green, blue–green, blue–violet, and red–violet.

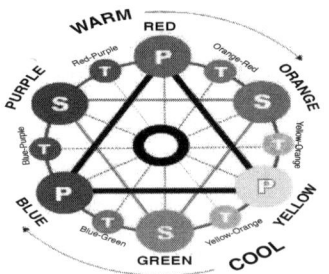

Prang colour chart

2. Munsell colour chart

Munsell colour system is a colour space that specifies colours based on three colour dimensions, hue, value (lightness), and chroma (colour purity or colourfulness). Several earlier colour order systems had placed colours into a three dimensional colour solid of one form or another, but Munsell was the first to separate hue, value, and chroma into perceptually uniform and independent dimensions, and was the first to systematically illustrate the colours in three dimensional spaces.

As Munsell explains:

Desire to fit a chosen contour, such as the pyramid, cone, cylinder or cube, coupled with a lack of proper tests, has led to many distorted statements of colour relations, and it becomes evident, when physical measurement of pigment values and chroma is studied, that no regular contour will serve. A colour is fully specified by listing the three numbers for hue, value, and chroma.

1. Hue: Each horizontal circle Munsell divided into five principal *hues*: Red, Yellow, Green, Blue, and Purple, along with five intermediate hues halfway between adjacent principal hues. Each of these 10 steps is then broken into 10 sub-steps, so that 100 hues are given integer values. Two colours of equal value and chroma, on opposite sides of a hue circle, are complementary colours, and mix additively to the neutral grey of the same value.

2. Value: Value, or lightness, varies vertically along the colour solid, from black (value 0) at the bottom, to white (value 10) at the top. Neutral greys lie along the vertical axis between black and white. Several colour solids before Munsell's plotted luminosity from black on the bottom to white on the top, with a grey gradient between them, but these systems neglected to keep perceptual lightness constant across horizontal slices. Instead, they plotted fully-saturated yellow (light), and fully saturated blue and purple (dark) along the equator.

3. Chroma: Chroma, measured radially from the centre of each slice, represents the "purity" of a colour, with lower chroma being less pure (more washed out, as in pastels). Note that there is no intrinsic upper limit to chroma. Different areas of the colour space have different maximal chroma co-ordinates. For instance light yellow colours have considerably more potential chroma than light Purples, due to the nature of the eye and the physics of colour stimuli. This led to a wide range of possible chroma levels—up to the high 30s for some hue-value combinations (though it is difficult or impossible to make physical objects in colours of such high chromas, and they cannot be reproduced on current computer displays).

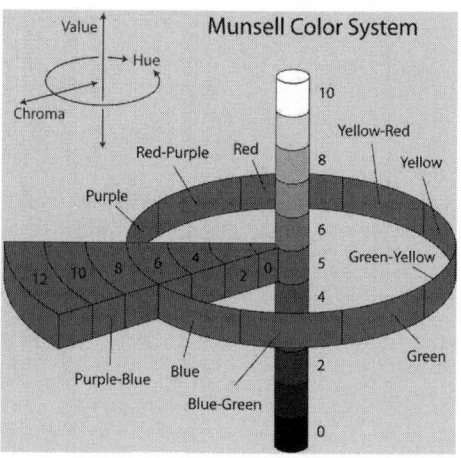

3. Physicist colour chart

A colour wheel based on RGB (red, green, blue) or RGV (red, green, violet) additive primaries has cyan, magenta, and yellow secondaries (cyan was previously known as cyan blue). Alternatively, the same arrangement of colours around a circle can be described as based on cyan, magenta, and yellow subtractive primaries, with red, green, and blue (or violet) being secondaries. Most colour wheels are based on three primary colours, three secondary colours, and the six intermediates formed by mixing a primary with a secondary, known as tertiary colours, for a total of 12 main divisions; some add more intermediates, for 24 named colours. Other colour wheels, however, are based on the four opponent colours, and may have four or eight main colours.

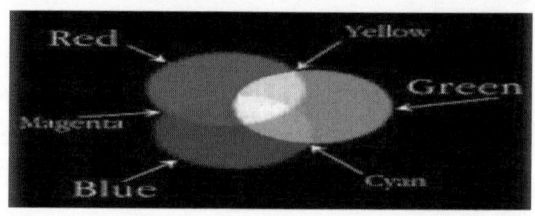

4. Psychology colour chart
Colour psychology: The colour white

- Purity
- Innocence
- Cleanliness
- Sense of space
- Neutrality
- Mourning (in some cultures/societies).

Colour psychology: The colour black

- Authority
- Power
- Strength
- Evil
- Intelligence
- Thinning/slimming
- Death or mourning.

Colour psychology: The colour grey

- Neutral
- Timeless
- Practical

Colour Psychology: The Colour red

- Love
- Romance
- Gentle
- Warmth
- Comfort
- Energy
- Excitement
- Intensity
- Life
- Blood

Colour psychology: The colour orange

- Happy
- Energetic
- Excitement
- Enthusiasm
- Warmth

- Wealth prosperity
- Sophistication
- Change
- Stimulation

Colour psychology: The colour yellow

- Happiness
- Laughter
- Cheery
- Warmth
- Optimism
- Hunger
- Intensity
- Frustration
- Anger
- Attention-getting

Colour psychology: The colour green

- Natural
- Cool
- Growth
- Money
- Health
- Envy
- Tranquillity
- Harmony
- Calmness
- Fertility

Colour psychology: The colour blue

- Calmness
- Serenity
- Cold
- Uncaring
- Wisdom
- Loyalty
- Truth
- Focused
- Un-appetising.

Colour psychology: The colour purple

- Royalty
- Wealth
- Sophistication
- Wisdom
- Exotic
- Spiritual
- Prosperity
- Respect
- Mystery

Colour psychology: The colour brown
- Reliability
- Stability
- Friendship
- Sadness
- Warmth
- Comfort
- Security
- Natural
- Organic
- Mourning (in some cultures/societies).

Colour psychology: The colour pink
- Romance
- Love
- Gentle
- Calming
- Agitation.

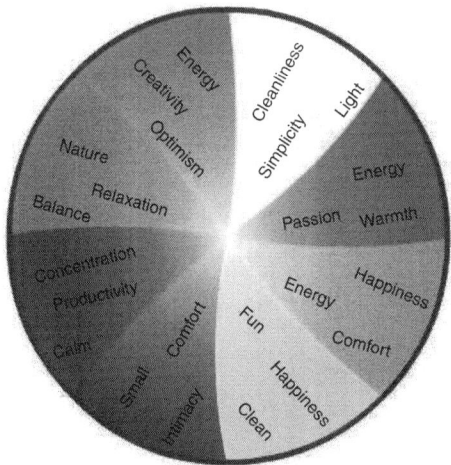

Colour Psycology

1.4   Design and construct garment for boy

Boy clothing is more casual. Nowadays a lot of kids wear are very much influenced by trends in adult wear. Good quality well-designed garments are a priority for a growing number of children's. Function and design must meet at the right proportions in boy clothing for it to be popular and accepted. Fabric choices, openings and fastenings, fit and ease, trimmings used are all major considerations when designing children's wear. Some other factors a designer designing for children's clothing should focus on are the changing shape of the growing kid and different proportions of the different parts of the body.

Design of the garment
Aim
To design and construct the garment for kid boy.

Features
Round neck, half sleeve, rib attached at neck, appliqué at front part, elastic attached at waist band, fly opening, patch pocket and set in pocket.
Components
T-shirt
> Front part cut–1 piece (pc)
> Back part cut–1 pc
> Sleeve part cut–2pcs
> Rib cut–1 pc

Shorts

> Front part cut–2 pcs
> Back part cut–2 pcs
> Waist band cut–2 pcs
> Fly opening cut–2 pcs (or) according to the taste

Materials required

Fabric, paper pattern and tool kit

Measurements details
T-shirt

Chest	:	21″
Neck	:	10½″
Full length	:	15″
Shoulder	:	6″

Shorts

Full length	:	23″
Inside leg	:	22″
Waist	:	21″
Seat	:	22″
Bottom	:	12″
Belt width	:	1½″

Drafting method

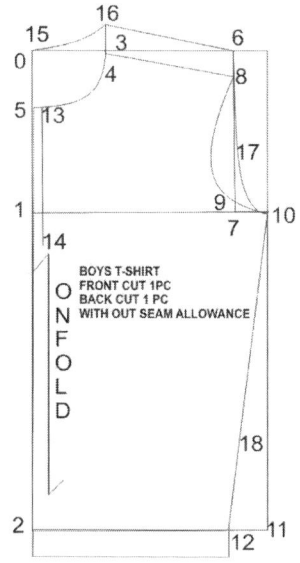
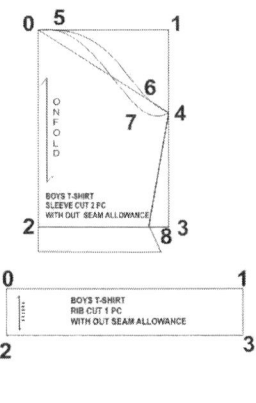

Front

Square lines from 0, fold at 0–2

1–0 = 1/4 chest

2–0 = full length plus 1 cm (1/4")

3–0 = 1/12 chest

4–3 = 2.5 cm (1")

5–0 = same as 3–0

Join 4–5, shape neck 5–3

6–0 = shoulder plus 1 cm (¼")

Square down from 6 to 7

8–6 = 2.5 cm (1"), Join 3–8

9–7 = about 2.5 cm (1")

10–1 = one-fourth chest plus 4 cm (1½")

Shape scye 8–9–10

Square down from 10 to 11

12–11 = 2.5 (1"). Join 10–12

Draw line 13–14 at a distance of 0.75 cm (¼") from line 5–1–2

14–13 = one fourth chest plus 2.5 cm (1") or to taste, for opening.

For this garment no opening for bodice front pattern.

Back

15–0 = 0.75 cm (¼")

16–3 = 3 cm (1¼")

Shape 15–16 and join 16–6

Shape syce 6–17–10

18–12 = nearly 10 cm (4"), notch for opening

Keeping 4 cm (1½") inturns at 2–12

Sleeve

0–1 = 1/4 chest

0–2 = sleeve length from shoulder plus 2 cm, (1½") for bottom folding

3–2 = same as 0–1, Join 3–1

4–1 = 1/8 chest less 1.25 cm (1/2")

0–5 = 4 cm (1½"), join 4–5

Shape back side 4–6–5–0 as shown

Square up from 4 to 7

7–4 = 1/12 chest

Taking 1 cm (1/4") above point 4

Shape front side 4–7–0 as shown

2–8 = 1/8 chest plus 6.5 cm (2½") or taste

Join 4–8 by straight line

Keep 1 or 4 cm, outside 2–8 for hem or inturns

Rib

0–1 and 2–3 = rib height

0–2 and 1–3 = ½ neck width

Rib part cut line from 0, 2, 3 and 1

Ribs can be also used in neck finishing in a single jersey fabric.

Shorts
Front

Square line from 0

1–0 = 1/4 seat plus 5 cm (2″) less belt width

2–0 = full length less belt width plus 1 cm (1/4″)

3–1 = 4 cm (1½″) for looseness

4–3 = 1/4 seat

5–0 = Same as 4–1. Join 4–5

6–5 = ¼ waist plus 7 cm (2¾″) for seam and pleats

7–4 = 1/6 seat

8–4 = 1/12 seat

Strike a line mid-way across the angle 7–4–8

9–4 = half 8–4 plus 0. 75 cm (1/4″). Shape fork 7–9–8

10–8 = 1.5 cm (1/2″)

11–2 = same as10–1, or half bottom round. Shape 8–11

12–5 = 1/12 seat

13 is the mid-way 12–6

Take 3.25 cm (1¼″) pleat at 12 and 2.5 cm (1″) pleat at 13

Pocket

14–6 = 3 cm (1¼″)

15–14 = 1/6 seat for pocket opening

Back part

16–8 = 4 cm (1½″)

17–11 = 3 cm (1¼″)

Shape 16–17

18–16 = 0.75 cm (¼″)

Join 10–7 and produce to 19–20

20–19 = 2.5–3 cm (1–1¼″) according to flat or prominent seat

21–20 = 1.5 cm (1/2″)

Join 21–7. Shape fork 7–22–18

23–21= 1/4 waist plus 4 cm (1½″). In this draft point 23 and 0 have coincided

24–23 = 1/12 seat plus 1.5 cm (1/2″)

Take a dart, 1.5 cm (1/2″) wide and 7.5 cm (3″) long at 24

Keep 5 cm (2″) below 2–11 and 2–17 for inturns

Fly

The fly to be attached at the centre front of shorts should be drafted as follows.

Use the front fork curve and draw the shape

4–1 = nearly 5 cm (2″)

5–2 = same as 4–1

6–3 = 1.5 cm (1/2″)

Shape 4–5–6–3–2–1

Drafting details for shorts

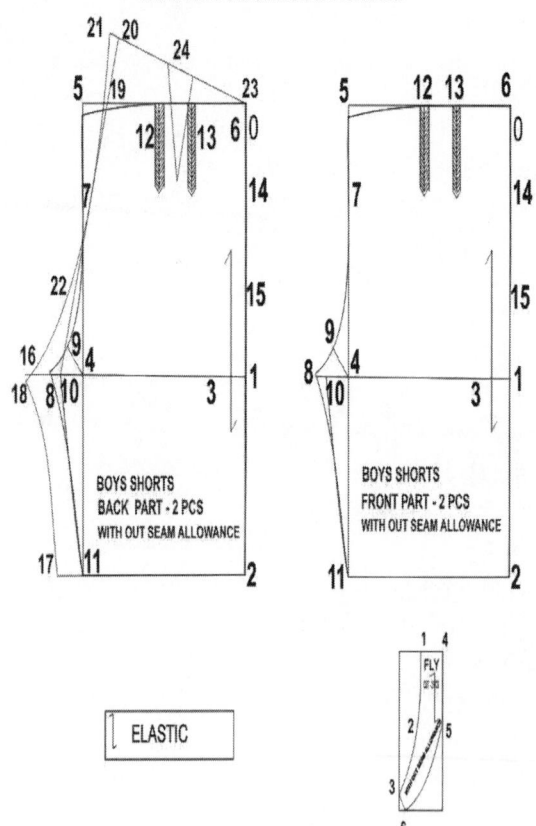

DRAFTING DETAILS FOR SHORTS

Fabric required for construct this garment

Approximately 2.00 m fabric required for this garment.

Layout—Combination layout

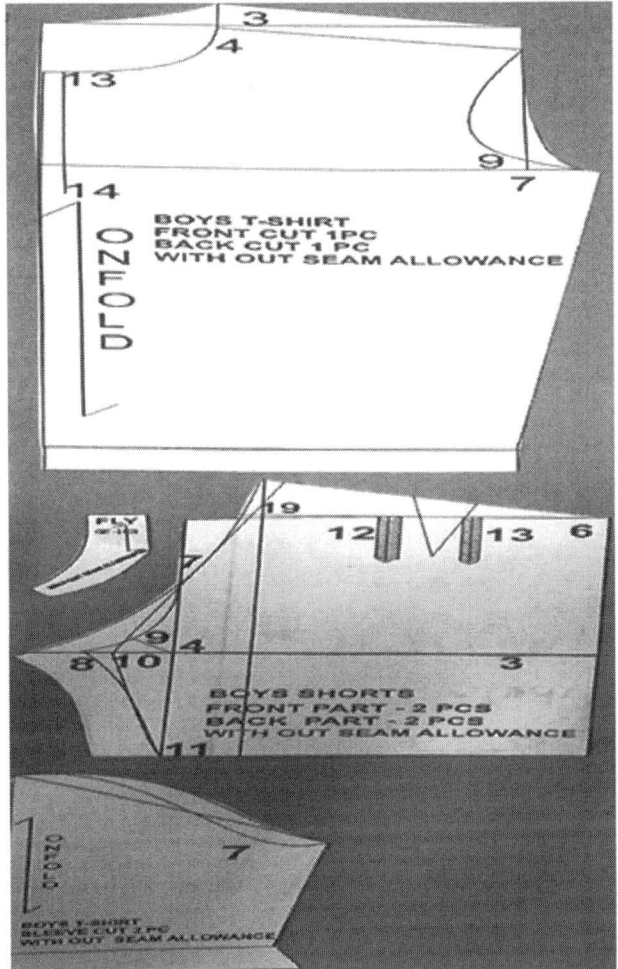

Construction procedure

T-shirt

The bodice part of the front and back part of the T-shirt is been constructed by finishing the neck line.

Join the shoulder part of front and back bodice.

Match the front and back bodice stitch the side seams.

Finish the bottom hem line.

Attach sleeve.

Shorts

Attach fly to the front part of the shorts.
Attach zippers throughout the length of the placket.
Join the front and back pieces to each other on right and left side.
Attach pocket to both sides of the pant at the back pieces.
Attach waist band and belt loops to it.
Finish the bottom hem line and press the garment well using seam line.

Trims and accessories
Zipper, elastic and appliqué.

Cost calculation
  i.   Cost of material: Rs 400
  ii.   Cost of accessories: Rs 50
  iii. Cost for construction: Rs 300
  iv.   Total cost of the garment: Rs 750

1.5   Design and construct garment for girl

Children from age group of 2–8 years are called toddlers. Fashion in clothing will be a reflector of change in life style .Children are given more care for the selection of their own garment. Innovative design and vivid colours have become the key for the fashioned wear. High fashion garments can be worn during special occasion like parties. While selecting the fabric for children choose durable and attractive fabrics. Skirt plays a major role in various girls' fashion garments. If a bodice is attached to a skirt, it becomes a frock.

DESIGN OF THE GARMENT

Aim

To design and construct the garment for children (girl).

Features

Round neck, gathers at waist line, sleeve less, embellishment at the waist line.

Components

Front bodice part cut 1 pc
Back bodice part cut 1 pc
Circular skirt part cut 1 pc
Inner skirt part cut 1 pc

Materials required

Fabric, paper pattern and tool kit

Measurements details

Chest	:	22″
Waist	:	22″
Full length'	:	24″
Shoulder	:	5″
Bodice length	:	10″
Belt width	:	1.5″

Method of drafting

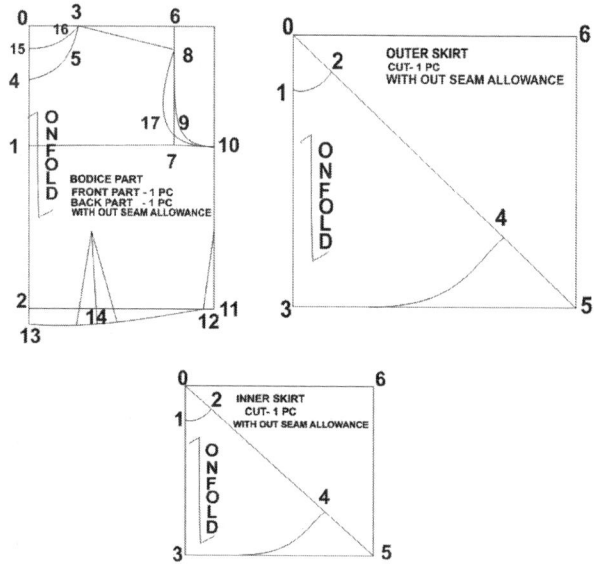

Front

Square lines from 0, on a four layer fold, with folds at 2–0 and 6–0

1–0 = 1/8 chest plus 6.5 cm (2½″)

2–0 = bodice length plus 1.5 cm (½″)

Square out from 1 and 2

3–0 = 1/12 chest, or plus 1 cm (1/4″) or to taste

4–0 = 1/8 chest or to taste, square out from 4 to 5

5–4 = 3–0 less 1 cm (1/4″), Join 3–5

6–0 = shoulder plus 1 cm (¼″)

Square down from 6 to 7

8–6 = 1.5 cm (½″). Join 8–3,

9–7 = about 2.5 cm (1″)

10–1 = 1/4 chest plus 4 cm (1½″)

Shape scye 8–9–10 as shown

Square down from 10 to 11

12–11 = 1.5 cm (½″)

Shape side seam 10–12

13–2 = 1.5 cm (½″)

Shape bottom 13–12

14–13 = 1/12 chest plus 1.5 cm (½″)

Take 1 cm (1/4″) dart at 14 for front and back

Back

15–0 = 2–2.5 cm (3/4″–1″)

Square out from 15 to 16

16–15 = same as 5–4

Join 16–3

Shape syce 8–17–10 as shown

Allow 2 cm (¾″) inlays at 12–10

Inner skirt

Square lines from 0

1–0 = 1/6 waist less 0.75 cm (1/4″)

Shape 1–2 with 0–1 radius (i.e., 2–0 is same as 1–0)

3–1 = required length less belt width plus 1 cm (1/4″)

Shape 3–4 with 0–3 radius,

Keep about 2 cm (3/4″) below 3–4, for inside turning

After cutting on lines 1–2 and 3–4, the unfolded cloth will look like a circle

Outer skirt

Square lines from 1–0 = one-sixth waist less 0.75 cm = (1/4″)

Shape 1–2 with 0–1 radius (i.e., 2–0 is same as 1–0)

3–1 = full length less belt width plus 1 cm (1/4″)

Shape 3–4 with 0–3 radius,

Keep about 2 cm (3/4″) below 3–4, for inside turning

After cutting on lines 1–2 and 3–4, the unfolded cloth will look like a circle

Fabric required for construct this garment

Silk cotton plain fabric, net fabric of same colour, lining material of same colour fabric, approximately 2.00–2.5 m fabric required for this garment.

Layout—Combination layout

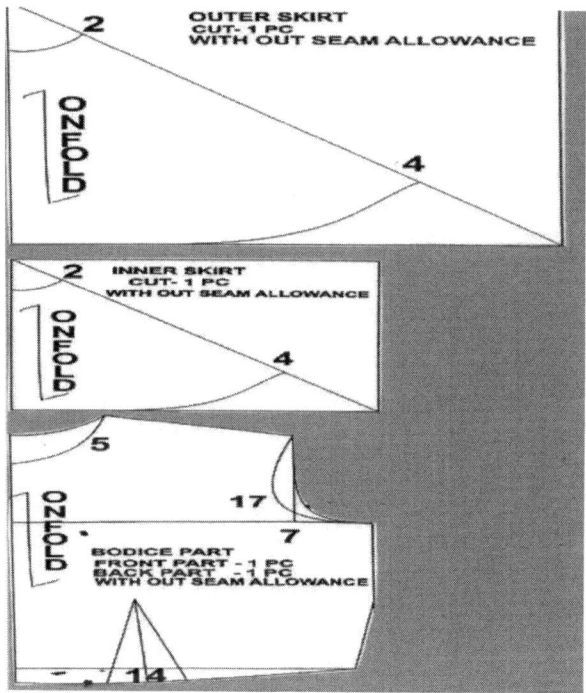

Construction procedure

- The bodice part of the frock is been constructed without sleeve.
- Join front and back shoulder pieces to each other on right and left side.
- In the back piece it is slashed in the centre till the required length of the placket for making placket (optional).
- Bias binding can be done to finish the neck and armscye depth of the garment.
- Match the front and back bodice and stitch the side seams.
- The bottom of circular skirt is been joined with its interlining fabric.
- Similar to circular skirt a netted fabric of larger size is attached to it at the top of the fabric.

- Both the bodice and skirt is been joined together.
- A waist band with a decorative flower is been attached at waist part.
- Finally hem the lower edge of the garment and press the garment well using steam line.
- Attach zippers throughout the length of the placket (optional).

Trims and accessories

Zipper, lace and bow

Cost calculation

   i.   Cost of material (fabric and lining): Rs 700
  ii.   Cost of accessories: Rs 100
 iii.   Cost for construction: Rs 300
 iv.   Total cost of the garment: Rs 1100

1.6   Design and construct garment for women

The anarkali suit is a form of women's dress. The anarkali suit is made up of long, frock-style top. The anarkali is an extremely elegant style that is worn by women. The anarkali suit varies in different lengths and embroideries of different colour combinations, including floor length anarkali styles. The kameez of the anarkali can be with sleeve or without sleeve. Many women preferred to wear anarkali suits for wedding purpose.

DESIGN OF THE GARMENT

Aim

To design and construct the garment for women

Features

High neck stand collar, full sleeve, front opening, embellishment at cuff and front neck line.

Components

Front bodice part cut–1 pc
Back bodice part cut–1 pc
Circular skirt part cut–1 pc
Full sleeve cut–2 pcs

Materials required

Fabric, pattern paper and tool kit

Measurements details

Chest	:	32"
Waist	:	25"
Full-length	:	14"
Shoulder	:	14.5"
Seat	:	34"
Sleeve with shoulder	:	23"
Sleeve round	:	6.5"
Skirt length	:	42"

Method of drafting

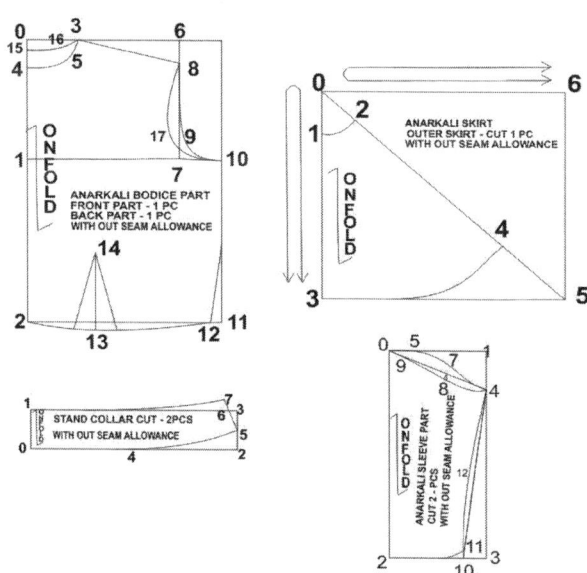

Front

Square lines from 0, on a four layer fold, with folds at 2–0 and 6–0

1–0 = 1/8 chest plus 6.5 cm (2½")

2–0 = bodice length plus 1.5 cm (½")

Square out from 1 and 2

3–0 = 1/12 chest, or plus 1 cm (1/4") or to taste

4–0 = 1/8 chest or to taste

Square out from 4 to 5

5–4 = 3–0 less 1 cm (1/4"), Join 3–5

6–0 = ½ shoulder plus 1 cm (¼")

Square down from 6–7

8–6 = 1.5 cm (½"). Join 8–3

9–7 = about 2.5 cm (1")

10–1 = 1/4 chest plus 4 cm (1½")

Shape scye 8–9–10 as shown

Square down from 10 to 11

12–11 = 1.5 cm (½")

Shape side seam 10–12

13–2 = 1.5 cm (½")

Shape bottom 13–12

14–13 = 1/12 chest plus 1.5 cm (½")

Take 1 cm (1/4") dart at 14 for front and back

Back

15–0 = 2–2.5 cm (3/4"–1")

Square out from 15 to 16

16–15 = same as 5–4

Join 16–3

Shape syce 8–17–10 as shown

Allow 2 cm (¾") inlays at 12–10

Skirt

Square lines from 0

1–0 = 1/6 waist less 0.75 cm

Shape 1–2 with 0–1 radius (i.e., 2–0 is same as 1–0)

3–1 = full length less belt width plus 1 cm (1/4")

Shape 3–4 with 0–3 radius

Keep about 2 cm (3/4") below 3–4, for inside turning

After cutting on lines 1–2 and 3–4, the unfolded cloth will look like a circle

Sleeve

Square line from 0, fold at 2–0

1–0 = one-eighth chest plus 6.5 cm (2.5")

2–0 = sleeve length plus 1.5 cm (1/2")

3–1 = same as 2–0. Join 2–3
4–1 = one-eighth chest less 1.5 cm (1/2")
5–0 = 2.5 cm (1"). Join 4–5
Shape 4–7–5–0 and 4–8–9–0 as shown
3–10 = 2"
Join 4–10. 11–10 = 1.5 cm (1/2") Shape 11–2 and 4–12–11 as shown

Collar

Square line from 0. Fold at 0–1
1–0 = 5 cm (2")
2–0 = Half neck plus 2 cm (3/4")
3–1 = Same as 2–0
4–0 and 5–2 = 1.5 cm (1/2")
6–2 = 1.5 cm (1/2")
7–3 = 2 cm (3/4"). Join 6–7. Shape 7–6–5–4–0 and 1 as shown in figure

Fabric required for construct this garment

Silk cotton plain and printed fabric of contrast colour. Lining material of same colour fabric, approximately 5 m material required for this design garment.

Layout—Combination Layout

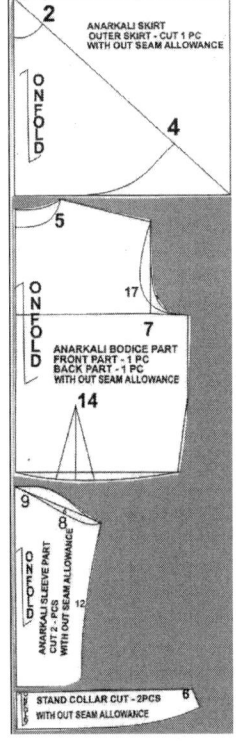

Construction details

- At front part, slit your required length of the bodice in centre front and zipper is attached at the centre part of the bodice.
- Place front and back bodice pieces by facing right sides together and sew.
- Attach the stand collar.
- Shoulder line to join the two front and back pieces together. Fold the hem line of sleeve by folding ¼″ inside and fold again ¾″ inside and sewing along the hemline at the edge of first folding.
- Place the full sleeve right side on the top of bodice armhole area, respectively. Sew along with ¼″ armhole seam allowance.
- Turn wrong side out and sew along with 1″ side seam allowances together.
- Complete the sleeves.

Trims and accessories

Zipper, embellishments at neck and cuff.

Cost calculation

i.	Cost of material	: Rs 1000
ii.	Cost of accessories	: Rs 550
iii.	Cost of construction	: Rs 800
iv.	Total cost	: Rs 2350

1.7   Design and construct garment for men

Shirt, a loose garment for the upper part of the body, is prepared in many styles and fashions, the popular and common type of shirts are full open shirts with full sleeves with closed collars, yoke and cuff, matched with bottom pant.

Aim

To design and construct the garment for men.

Features

Shirt front opening, collar, full sleeve, yoke attached at the front bodice, slim fit trouser, zipper fly opening, waistband, etc.

DESIGN OF THE GARMENT

Components
Shirt
>Front part cut = 2 pcs
>Back part cut = 1 pc
>Back yoke cut = 2 pcs
>Full sleeve cut = 2 pcs
>Shirt collar cut = 2 pcs

Pant
>Front part cut = 2 pcs
>Back part cut = 2 pcs
>Waistband cut = 2 pcs
>Fly-opening cut = 2 pcs

Materials required
Fabric, pattern paper and tool kit

Measurements details
Shirt
>Chest   :   34″
>Neck   :   14 ¾″

Shirt length   :   32"
Shoulder   :   8.5"
Full sleeve length   :   24"

Pant

Full length   :   44"
Inside leg   :   30"
Waist   :   32"
Seat   :   36"
Bottom   :   17"
Belt width   :   11/2"

Method of drafting

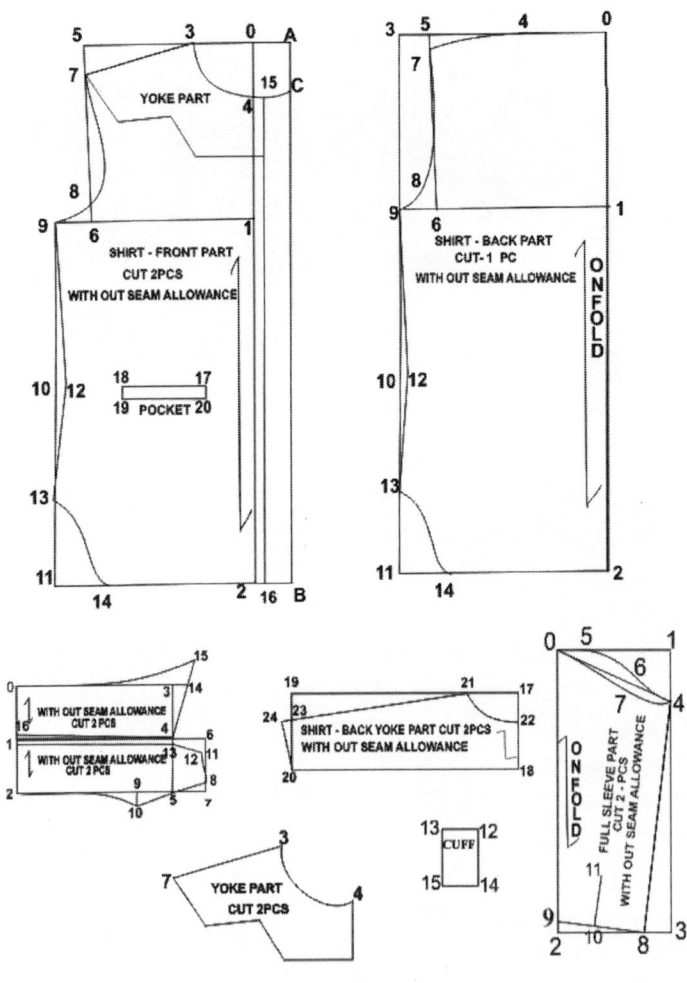

Front

Draw line 0–1–2 at a distance of 4 cm (1½″) from the fold A–B.

    1–0 = one-fourth chest

    2–0 = full length plus 2.5 cm (1″)

    Square out from 0, 1 and 2

    3–0 = one-sixth neck

    4–0 = one-sixth neck plus 0.75 cm (1/4″)

    Shape neck 4–3 as shown

    5–0 = shoulder plus 0.75 cm (1/4″)

    6–1 = one-fourth chest. Join 5–6

    7–5 = one-fourth of 5–6

    Join shoulder 3–7

    8–6 = about 3 cm (11/4″)

    9–1 = one-fourth chest plus 5 cm (2″) or to taste

    Shape scye 7–8–9

    Square down from 9 to 10 and 11

    10–9 = about one-fourth chest

    12–10 = 1.5 cm (1/2″)

    13 is mid-way 10–11

    14–11 = about one-eighth chest

    Shape 9–12–13–14 as shown

    15–4 = 1.5 cm (1/2″)

    16–15 (parallel to 1–0) = one-fourth chest plus 10 cm (4″)

Cut the lower layer from 15 to 16 of the right side i.e., the left side should be wider than the right side

    C–A = same as 3–0, join 4–C

Pocket

    18–19 = about 5 cm (2″)

    17–20 = same as 18–19

    18–17 = one-eighth chest or plus 1.5 cm (1/2″)

    Mark pocket as shown.

Back

Cut the front and use it for back in such a way that lines 1–2 and 1–9 of front and back will coincide. The back is cut in the following two ways.

    Plain back without pleats or gathering

    Square lines from 0, fold at 0–2

    1–0 = same as 1–0 front less 4 cm (11/2″)

    2–1 = same as 2–1 front

    Square out from 0, 1 and 2

    5–0 = same as 5–0 of front

    Square down from 5 to 6

7–5 = 1.5 cm (1/2″)

4 is mid-way 5–0

Shape 7–4 as shown

Square lines from 17, fold at 17–18

18–17 = one-eighth chest

19–17 = shoulder plus 0.75 cm (1/4″)

20–18 = same as 19–17, join 20–19

21–17 = one-sixth neck plus 0.75 cm (1/4″)

22–17 = half of 21–17 plus 1 cm (1/4″)

Shape neck 22–21

23–19 = 3 cm (1¼″) for boys

Keep 21–23 and produce 0.75 cm (1/4″) to 24 Join 24–20

Sleeve

Square lines from 0, fold at 0–2

1–0 = one-fourth chest less 1.5 cm (1/2″)

2–0 = sleeve length from shoulder, plus 2 cm (3/4″) for seams

3–2 = same as 1–0. Join 3–1, 4–1 = one-eighth chest less 1.25 cm (½″)

5–0 = 4 cm (1 ½″)

Join 4–5

Shape back side 4–6–5–0 as shown

Square up from 4 to 7

7–4 = one-twelfth chest

Taking 1 cm (1/4″) above point 4

Shape front side 4–7–0 as shown

8–2 = one-eight chest plus 6.5 cm (2 ½″), or to taste

Join 4–8 by straight line

Keep 1 or 4 cm (1/4 or 1 ½″) out-side 2–8 for hem or inturns

9–2 = 1.25 cm (1/2″). Join 9–8

10–8 = half of 8–9 plus 1.5 cm (1/2″)

11–10 = one-sixth chest,

5–6 = cuff opening

Collar

Square lines from 0, fold at 0–2

1–0 = 6 cm (21/4″)

2–1 = 4 cm (11/2″)

3–0 = half neck plus 1 cm (1/4″)

4–3 = same as 1–0

5–4 = same as 2–1

6–4 = 4 cm (11/2″)

7–6 = same as 2–1

8–7 = 1 cm (1/4″)

9–5 = 5 cm (2″)

10–9 = 1 cm (1/4″), shape 8–5–10–2 as shown

11–6 = 1.5 cm (1/2″)

12–11 = 0.75 cm (1/4″). Join 12–8

13–4 = 1 cm (1/4″)

Join 13–14 and produce the line to 15

15–13 = 10–13 cm (4–5″), shape 15–0 as shown

16–1 = 1 cm, shape 16–13–12

Cuff

Length = one-fourth chest plus 1.5 cm (1/2″) or to taste

Width = one-sixteenth chest plus 1.5 cm (1/2″) or to taste

Single cuff

Square the lines 12 fold at 13–12

13–12 = cuff width plus 1.25 cm (½″) or nearly 7.5 cm (3″)

14–12 = half cuff length plus 1 cm (¼″), or one-eighth chest plus 1 cm (¼″)

15–13 = same as 14–12 or less 0.75 cm (¼″)

Join 15–14

Method of drafting
Pant

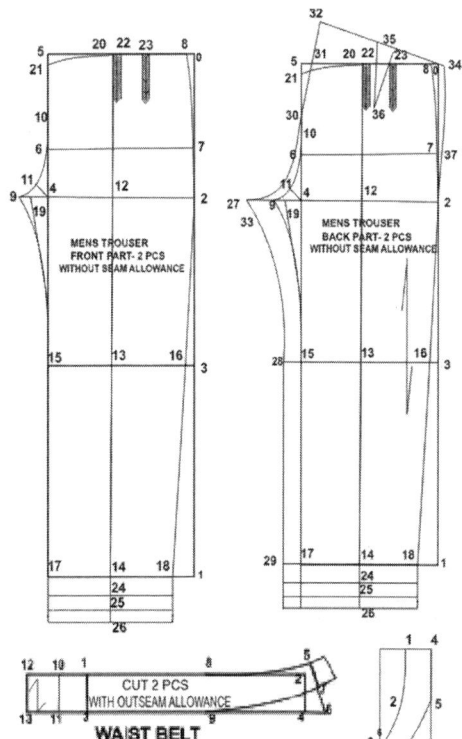

Front

1–0 = full length less belt width plus 1 cm (1/4″)

2–1 = inside leg plus 1 cm (1/4″)

3–2 = half inside leg less 5 cm (2″)

Square lines from all these points

4–2 = one-fourth seat plus 4 cm (11/2″)

5–0 = same as 4–2. Join 4–5

6–4 = one-twelth seat

7–2 = same as 6–4. Join 6–7

8–5 = one-fourth waist plus 7.5 cm

(3″) for pleats and seams, shape 8–7 as shown

9–4 = one-twelth seat less 1.5 cm (1/2″)

10–4 = one-eighth seat plus 1.5 cm (1/2″)

11–4 = half of 9–4 plus 0.75 cm (1/4″), shape fork 10–11–9

12 is mid-way 9–2

Square down from 12–13–14

17–14 = one-fourth bottom

18–14 = same as 17–14

19–9 = 2 cm (3/4″)

Join 19–17 and shape 9–15

16–13 = same as 15–13

Join 16–18 and shape 7–16

20 is squared up from 12

21–5 = 1 cm (1/4″). Shape 20–21

22–20 = 4 cm (11/2″), for pleat

23 is mid-way 22–8

Suppress 2.5 cm (1″) in the pleat at 23

24–14 and 25–24 = each 4 cm (11/2″), width of p.t.ups

26–25 = 2.5 cm (1″)

Draw lines parallel to 17–18 from points 24, 25 and 26

Mark pocket at a distance of 3 cm (11/4″) below 8

Pocket opening = one-sixth seat

Back

27–9 = 5–6.5 cm(2–21/2″)

28–15 = 2.5 cm (1″)

29–17 = 2.5 cm (1″)

Shape 27–28 and join 28–29

30–10 = 2.5 cm (1″) for seat angle

This quantity should be increased for an erect figure and reduced for a stopping figure.

Join 19–30 and produce to 31–32

32–31 = 2.5–4 cm (1–11/2″) according to flat or prominent seat

33–27 = 1 cm (1/4″), shape fork 30–33 as shown

Draw side seam 34–16–18

35–34 = one-twelfth seat plus 1.5 cm (1/2″)

36 is squared down from 35 and equals 9 cm (31/2″)

Add 10 to 11.5 cm (4–4 ½″), below 18–29 for p.t.ups

Fly

The fly to be attached at the centre front of pants should be drafted as follows:

Use the front fork curve and draw the shape

4–1 = nearly 5 cm (2″)

5–2 = same as 4–1

6–3 = 1.5 cm (1/2″), shape 4–5–6

Waist belt

2–1 = half waist plus 1 cm (1/4″)

3–1 = 3.25 cm (11/4″) or to taste

4–2 = same as 3–1. Join 3–4

5–2 = 1.25 cm (1/2″)

6–4 = same as 5–2. Join 5–6

7–5 = same as 3–1 or less 1 cm (1/4″)

8–2 and 9–4 = one-sixth waist, shape 5–8 and 7–9

Draw 10–11 at a distance of 4 cm (11/2″) from 1–3, the extension for the right side

If an extended belt is desired, extend the left side to 12–13 which is 8–12.5 cm (3–5″) from 1–3

Keep about 4 cm (11/2″) inlays outside 5–6, see cross-line

Fabric required for construct this garment

Approximately 5.5 m fabric required for this garment.

Layout—Lengthwise fold

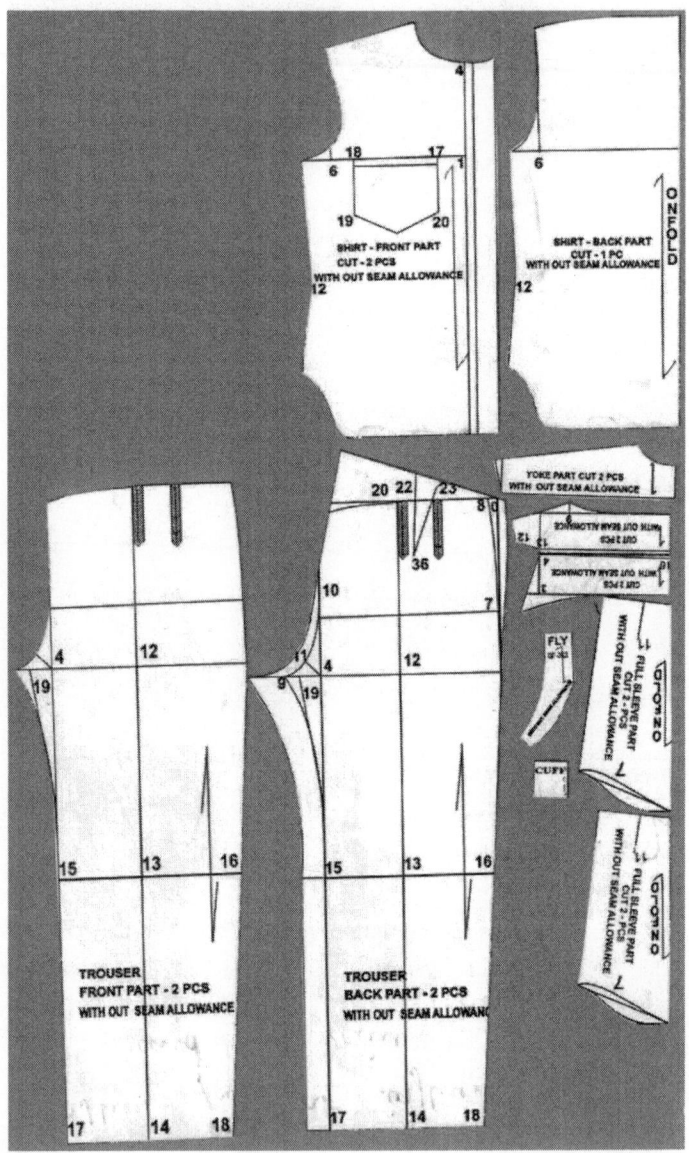

Construction procedure

In the front bodice attach placket.

Attach buttons.

Join shoulder part of the back and front bodice.

Attach collar piece to the bodice and finish it.

Join the front and back pieces to each other on right and left sides.

Attach the cuff to the sleeve bottom.

Join sleeve to the bodice.

Join the front and back pieces to each other on right and left sides.

Finish the bottom hem line and press the garment well using seam line.

Pant

Attach fly to the front part of the pant.

Attach zippers throughout the length of the placket.

Attach pocket to both sides of the pant at the back pieces.

Join the front and back pieces to each other on right and left side.

Attach waist band and belt loops to it.

Finish the bottom hem line and press the garment well using steam line.

Trims and accessories

Zipper, hooks and buttons.

Cost calculation

  i.   Cost of material: Rs 1000
 ii.   Cost of accessories: Rs 50
iii.   Cost of construction: Rs 800
 iv.   Total cost: Rs 1850

1.8   References

1. Tomoko Nakamichi (2011), Pattern magic 2, Laurence King Publishing Ltd, 361-373 City road, London EC1V1LR, United Kingdom, ISBN 978-1-85669-706-4.

2. Winifred Aldrich (2015), Metric pattern cutting for women's wear, 6th edition, Wiley, ISBN 978-1-119-02828-4 (ebk).

3. Anita Tyagi. Garment Construction. Sonali Publications, New Delhi-110002, ISBN 978-81-8411-487-4.

4. Aftab Ahmed Khan. Garment Technology, Random Publications, New Delhi-110001, ISBN 978-93-5111-098-9.

5. Winifred Aldrich (2015), Metric pattern cutting for men's wear 5th edition, Reprint 2015, Wiley India Pvt Ltd, 4435-36/7, Ansari Road, Daryaganj, New Delhi-110002, ISBN 978-81-265-3241.

6. Anita Tyagi. Couture Sewing Techniques, Sonali Publications, 4228/1, Ansari Road, New Delhi-110001, India, p 190, ISBN 978-81-8411-532-9.

7. Zoya Nudelman (2016), The art of couture sewing, 2nd edition, Chicago, IL, Bloomsbury Publishing Inc. ISBN 978-1-60901-831-3.

8. Mary Mathews (1985), Practical Clothing Construction, Part 1, Basic Sewing Process. Bhattarams w8 VSI Estate, Thiruvanmiyar, Chennai-600034.

9. Zarapkar Tailoring College. Zarapkar System of Cutting. Navneet Publications (India) Limited, Dantali, Gujarat. Printed by Shreeji offset, 99, Amrut Industrial

Estate, Ahmadabad, 380052. ISBN 81-243-0199-9. pp 40, 59, 67, 75, 76, 125, 126, 149, 152, 157, 159, 160 and 163.

10. Basic pattern development, p 25–26, students hand book, practical manual class XII.

11. Indian Garment Design Course book, edition 2011, copyright 2011, Usha International Limited.

12. Standers of quality control, pp 17–21, published by Association of Sewing and Design Professionals, copyright@2008. www.sewing professionls.org.

13. Yoke - Definition and More from the Free Merriam-Webster Dictionary. Merriam-Webster.com. Retrieved 8 June 2012.

14. Cumming, Valerie, C.W. Cunnington, P.E. Cunnington (2010). The Dictionary of Fashion History (revised ed.). Berg. p 227. ISBN 1847885330.

15. Henley Shirt. H&M. Archived from the original on 2016-03-23. Retrieved 2016-06-04.

16. Homespun Knitwear coalminer contrast Henley. Archived from the original on 2014-08-22. Retrieved 2016-06-04.

17. Indigo Tissue Long Sleeve Henley. American Apparel. Archived from the original on 2016-06-04. Retrieved 2016-06-04.

18. www.textileschool.com.

19. www.Artvictus.com.

20. www.craftsy.com/blog/2014/02/types of darts in sewing.

21. https://artofstyle.club/yokes-on-dress-shirts.

22. Technical source book for designs by Jacil Lee, Camille Steen.

2.1   Design and construct uniform clothing for policeman

A **shirt** is a cloth garment for the upper body (from the neck to the waist). The dress—shirt is a button-up shirt with a collar and long sleeves. It is cut differently than the sport shirt, which is made to be worn open-necked. The dress shirt is designed to carry a jacket and tie, but can be worn without one or the other, or with neither.

    Trousers are worn on the lower part of the body from the waist to the ankles covering both legs separately. The Indian trouser had a narrow fall front, button closing front. Men commonly used narrow fall front with side pockets, pocket in the waistband and adjustable gusset at centre back.

Design of the garment
Aim
To design and construct the police uniform male.
Materials required
Fabric, pattern paper and tool kit.

Measurements details

Neck   :   35 cm (14")
Chest   :   80 cm (32")
Full length :   76 cm (30")
Shoulder   :   20 cm (8")
Full sleeve :   58 cm (23")

Method of drafting
Shirt

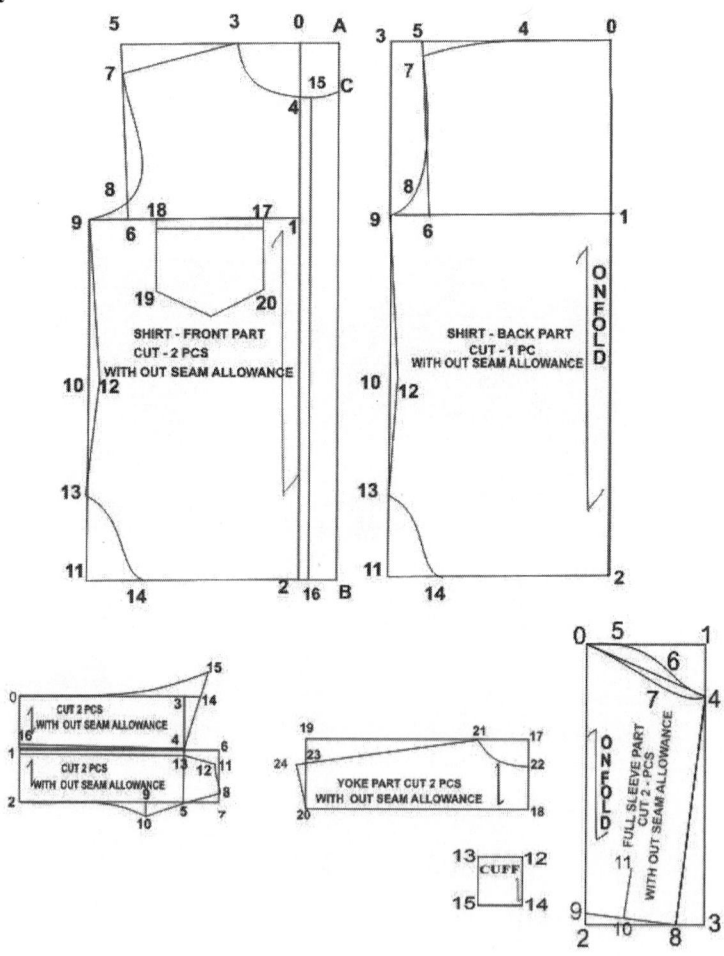

Front

Draw a line 0–1–2 at distance of 4 cm (1½") from the fold A–B
1–0 = one-fourth chest
2–0 = full length plus 2.5 cm (1"), Square out from 0, 1 and 2
3–0 = one-sixth neck

4–0 = one-sixth neck plus 0.75 cm (1/4"), Shape neck 4–3 as shown

5–0 = shoulder plus 0.75 cm (1/4")

6–1 = one-fourth chest join 5–6

7–5 = one-fourth of 5–6, join shoulder 3–7

8–6 = about 3 cm (1¼")

9–1 = one-fourth chest pulse 5 cm (2") or taste. Shape scye 7–8–9

Square down from 9 to 10 and 11

10–9 = about one-fourth chest

12–10 = 1.5 cm (1½")

13 is mid-way 10–11

14–11 = about one-eighth chest, Shape 9–12–13–14 as shown

15–4 = 1.5 cm (1½")

16–15 = (parallel to 1–0) = one-fourth chest plus 10 cm (4")

Cut the lower layer from 15–16 of the right side i.e., the left side should be wider than the right side.

C–A = same as 3–0. Join 4–C

Pocket

17–1 = about 5 cm (2")

18–17 = 1–8 chest or plus 1.5 cm (1/2"), width

Pocket height, same as 18–17 plus 1.5 cm (1/2")

18–19 and 17–20 is pocket height

Mark pocket as shown

Back

Cut the front and use it for back in such a way that lines 1–2 and 1–9 of front and back will be coincide. The back is cut as plain back without pleats or gatherings.

Square lines from 0, fold at 0–2

1–0 = same as 1–0 of front less 4 cm (1½")

2–1 = same as 2–1 of front, Square out from 0, 1 and 2

5–0 = same as 5–0 of front. Square down from 5–6

7–5 = 1.5 cm (½")

4 is mid-way 5–0, Shape 7–4 as shown

Except 8, the proportions of 9–1 are the same like front

Shape scye 7–8–9

Yoke

Square line from 17, fold at 17–18

18–17 = one-eight chest

19–17 = shoulder plus 0.75 cm (1¼")

20–18 = same as 19–17

Join 20–19

21–17 = one-sixth neck plus 0.75 cm (¼″)
22–17 = half of 21–17 plus 1 cm (¼″)
Shape neck 22–21
23–19 = 3 cm (1¼″) for boys keep it 2.5 cm (1″)
Join 21–23 and procedure 0.75 cm (1/4″) to 24. Join 24–20

Plain sleeve

Square lines from 0, fold at 0–2
1–0 = one-fourth chest less 1.5 cm (1/2″)
2–0 = sleeve length from shoulder, plus 2 cm (3/4″) for seams
3–2 = same as 1–0. Join 3–1
4–1 = one-eighth chest less 1.25 cm (½″)
5–0 = 4 cm (1 ½″)
Join 4–5
Shape back side 4–6–5–0 as shown
Square up from 4 to 7
7–4 = one-twelfth chest
Taking 1 cm (1/4″) above point 4
Shape front side 4–7–0 as shown
8–2 = one-eight chest plus 6.5 cm (2 ½″), or to taste
Join 4–8 by straight line
Keep 1 or 4 cm (1/4 or 1½″) out-side 2–8 for hem or inturns
9–2 = 1.25 cm (1/2″). Join 9–8
10–8 = half of 8–9 plus 1.25 cm (1/2″)
11–10 = one sixth chest, cuff opening = 5–6

Shirt collar

Square from 0, fold at 0–2
1–0 = 6 cm (2¼″)
2–1 = 4 cm (1½″)
3–0 = half neck plus 1 cm (¼″)
4–3 = same as 1–0
5–4 = same as 2–1
6–4 = 4 cm (1½″)
7–6 = same as 2–1
8–7 = 1 cm (¼″)
9–5 = 5 cm (2″)
10–9 = 1 cm (¼″) shape 8–5–10–2 as shown
11–6 = 1.5 cm (½″)
12–11 = 0.75 cm (1/4″) join 12–8
13–4 = 1 cm (¼″) shape 12–13–1
14–3 = 2.5 cm (1″) join 13–14 and produce the line to 15
15–13 = 10–13 (4–5″) or to taste

Shape 15–0 as shown. 16–1 = 1 cm (¼″)

Shape 16–13–12

Note: After cutting the material at 1–13–16 (see the cross-lines), we get two separate parts as follows:

Fall: 0–15–13–16–0

Stand: 1–12–8–10–2–1

Measurements details

Pant

Full length	:	104 cm (41″)
Inside leg	:	73 cm (29″)
Waist	:	76 cm (30″)
Seat	:	92 cm (36″)
Bottom	:	43 cm (17″)
Belt width	:	4 cm (1½″)

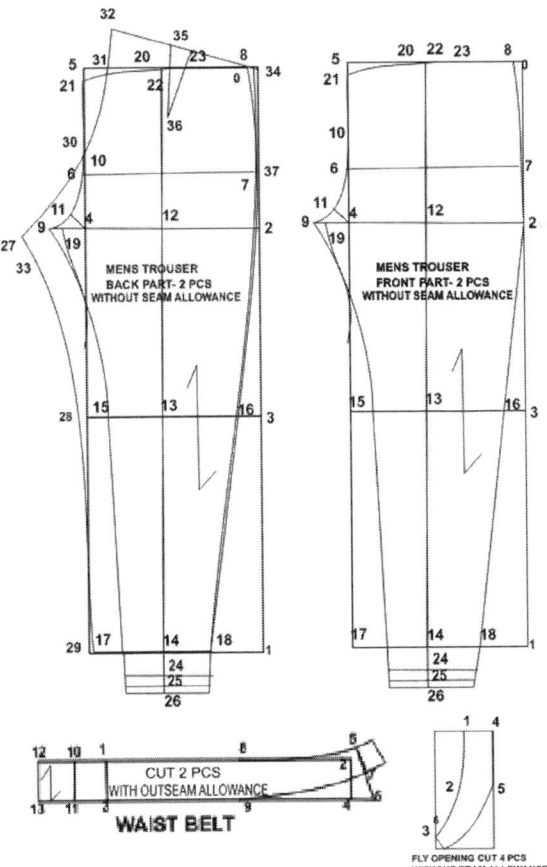

Front

Square lines from 0

1–0 = Full-length, less belt-width plus 1 cm (¼″)

2–1 = inside leg plus 1 cm (¼″)

3–2 = half inside leg less 5 cm (2″)

Square lines from all these points

4–2 = one-fourth seat pulse 4 cm (1½″)

5–0 = same as 4–2. Join 4–5

6–4 = one-twelfth seat

7–2 = same as 6–4. Join 6–7

8–5 = one-fourth waist plus 7.5 (3″) for pleats and seams shape 8–7 as shown

9–4 =one-twelfth seat less 1.5 cm (½″)

10–4= one-eight seat pulse 1.5 cm (½″)

11–4 half of 9–4 plus 0.75 cm (¼″)

Shape fork 10–11–9

12 is mid-way 9–2

Square down from 12–13–14

17–14 one-fourth bottom

18–14 same as 17–14

19–9 = 2 cm (¾″)

Join 19–17 and shape 9–15

16–13 = same as 15–13

Join 16–18 and shape 7–16

20 is squared up from 12

21–5 = 1 cm (¼″) shape 20–21

22–20 = 4 cm (1½″) for pleat, 23 is mid-way 22–8

Suppress 2.5 cm (1″) in the pleat at 23

24–14 and 25–24 = each 4 cm 1 (1½″)

26–25 = 2.5 cm (1″)

Draw lines parallel to 17–18 from points 24, 25 and 26

Mark pock at a distance of 3 cm (1¼″) below 8

Pocket opening = 1/6 seat

Back

27–9 = 5–6.5 cm (2–2½″)

28–15 = 2.5 cm (1″)

29–17 = 2.5 cm (1″)

Shape 27–28 and join 28–29

30–10 = 2.5 cm (1″) for seat angle. This quality should be increased for an erect figure and reduced for a stooping figure

Join 19–30 and produce to 31–32

32–31 = 2.5–4 cm (1–1½″) according to flat or prominent seat

33–27 = 1 cm (1/4″) shape fork 30–33 as shown

34–32 = one-fourth waist plus 4 cm (1½″)

Draw side-seam 34–16–18

35–34 = one-twelfth seat plus 1.5 cm (½″)

36 is squared down from 35 and equals 9 cm (3½″)

Add 10–11.5 cm (4–4½″)

Below 18–29 for p.t.ups

Waist belt

2–1 = Half waist plus 1 cm (1/4″)

3–1 = 3.25 cm (1¼″) or taste

4–2 = same as 3–1. Join 3–4

5–2 = 1.25 cm (½″)

6–4 = same as 5–2. Join 5–6

7–5 = same as 3–1 or less 1 cm (¼″)

8–2 and 9–4 = one-sixth waist

Shape 5–8 and 7--9

Draw 10–11 at a distance of 4 cm (1½″) from 1–3, the extension for the right side

If an extended belt is desired extend the left side to 12–13 which is 8–12.5 cm (3–5″)

Keep about 4 cm (1½″) inlays outside 5–6, see cross-line

Fly-opening

1–4 = 5 cm (2″)

5–2 = 5 cm (2″)

6–3 = 1.5 (1/2″), Shape 4–5–6

Fabric required for construct this garment

Khaki (Brown) colour cotton plain fabric, lining material of same colour fabric, approximately 5.5 m fabric required for this garment.

Layout—Combination layout

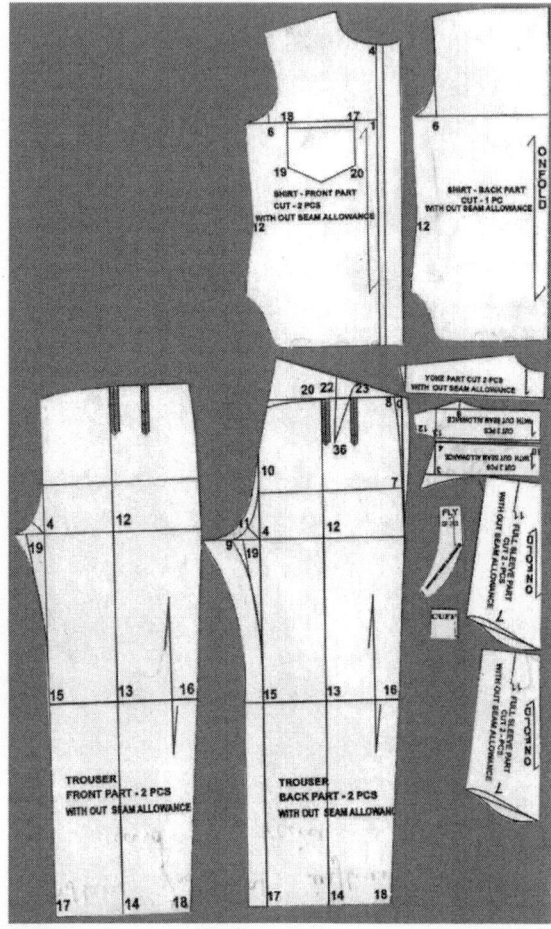

Construction procedure

- Joining the yoke part to back body part.
- Front side finish with plackets.
- Front right side finishes ¾″ for button fixing.
- Attach the pockets with flap at front left and right side.
- Join front and back pieces to each other on right and left side.
- Finishes the sleeves bottom hem.
- Joining the sleeves on armhole shapes.
- Match the front and back bodice and stitch the side seams.
- Complete bottom hem finish, attach a collar stand and collar with canvas finishing.
- Finishing back side dart, attach zippers front placket.

- Attach the side and back pockets.
- Finish the crotch point, joining the side seam and in leg seam.
- Finish the trouser bottom hem.
- Joint waist band to trouser part with canvas finishing.
- Fixing accessories and ironing.

Trims and accessories
Canvas, buttons, zipper, hook and eyelet.

Cost calculation
  i.   Cost of material: Rs 600
 ii.   Cost of accessories: Rs 120
iii.   Cost for construction: Rs 900
 iv.   Total cost: Rs 1620

2.2   Design and construct uniform clothing for army

A **military uniform** is the standardised dress worn by members of the armed forces of various nations. Military dress and military styles have gone through great changes over the centuries from colourful and elaborate to extremely utilitarian. Military uniforms in the form of standardised and distinctive dress, intended for identification and display, are typically a sign of organised military forces equipped by a central authority.

A **shirt** is a cloth garment for the upper body (from the neck to the waist). The dress shirt is a button-up shirt with a collar and long sleeves. **Trousers** are worn on the lower part of the body from the waist to the ankles covering both legs separately.

Design of the garment
Aim
To design and construct the army uniform.

Materials required
Fabric, pattern paper and tool kit

Measurements details

Neck   :   14"
Chest   :   32"
Full length :   29"
Shoulder   :   8"
Full sleeve :   22"

Method of drafting
Shirt

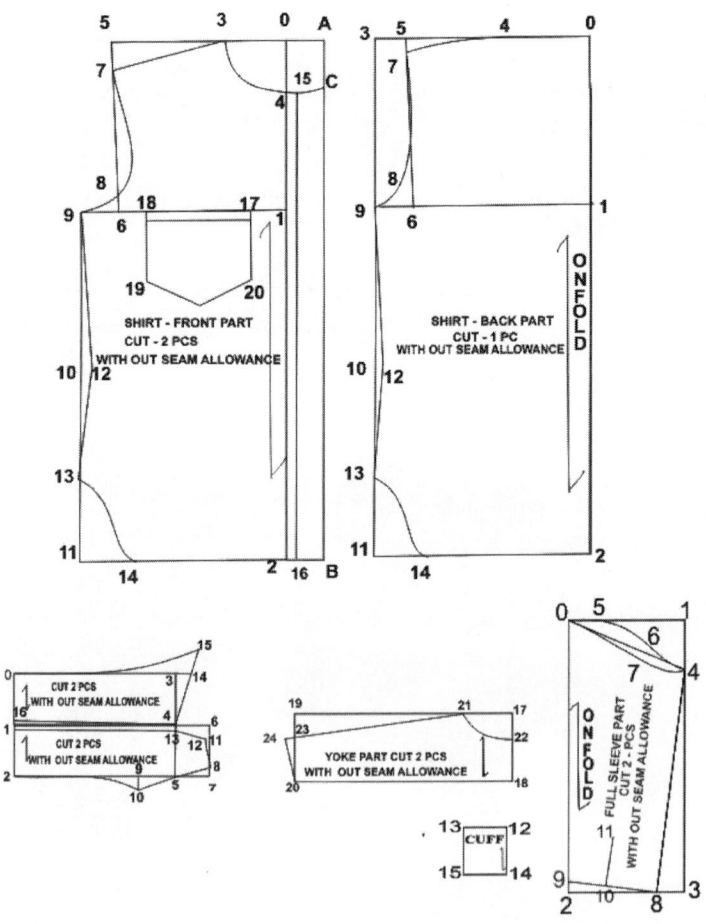

Front

Draw a line 0–1–2 at distance of 4 cm (1½") from the fold A–B

1–0 = one-fourth chest

2–0 = full length plus 2.5 cm (1") Square out from 0, 1 and 2

3–0 = one-sixth neck

4–0 = one-sixth neck plus 0.75 cm (1/4") Shape neck 4–3 as shown

5–0 = shoulder plus plus 0.75 cm (1/4")

6–1 = one-fourth chest join 5–6

7–5 = one-fourth of 5–6 join shoulder 3–7

8–6 = about 3 cm (1¼")

9–1 = one-fourth chest plus 5 cm (2") or taste

Shape scye 7–8–9

Square down from 9 to 10 and 11

10–9 = about one-fourth chest

12–10 = 1.5 cm (1½")

13 is mid-way 10–11

14–11 = about one-eighth chest

Shape 9–12–13–14 as shown

15–4 = 1.5 cm (1½")

16–15 (parallel to 1–0) = one-fourth chest plus 10 cm (4")

Cut the lower layer from 15 to 16 of the right side i.e., the left side should be wider than the right side.

C–A = same as 3–0. Join 4–C.

Back

Cut the front and use it for back in such a way that lines 1–2 and 1–9 of front and back will be coincide. The back is cut without pleats or gatherings.

Square lines from 0, fold at 0–2

1–0 = same as 1–0 of front less 4 cm (1½")

2–1 = same as 2–1 of front, Square out from 0, 1 and 2

5–0 same as 5–0 of front. Square down from 5–6

7–5 = 1.5 cm (½")

4 is mid-way 5 to 0, Shape 7–4 as shown

Except 8, the proportions of 9–1 are the same like front

Shape scye 7–8–9

Yoke

Square line from 17, fold at 17–18

18–17 = one-eight chest

19–17 = shoulder plus 0.75 cm (1¼″)

20–18 = same as 19–17

Join 20–19

21–17 = one-sixth neck plus 0.75 cm (¼″)

22–17 = half of 21–17 plus 1 cm (¼″)

Shape neck 22–21

23–19 = 3 cm (1¼″) for boys keep it 2.5 cm (1″)

Join 21–23 and procedure 0.75 cm (1/4″) to 24. Join 24–20

Plain sleeve

Square lines from 0, fold at 0–2

1–0 = one-fourth chest less 1.5 cm (1/2″)

2–0 = sleeve length from shoulder, plus 2 cm (3/4″) for seams

3–2 = same as 1–0. Join 3–1

4–1 = one-eighth chest less 1.25 cm (½″)

5–0 = 4 cm (1½″)

Join 4–5

Shape back side 4–6–5–0 as shown

Square up from 4 to 7

7–4 = one-twelfth chest

Taking 1 cm (1/4″) above point 4

Shape front side 4–7–0 as shown

8–2 = one-eight chest plus 6.5 cm (2½″), or to taste

Join 4–8 by straight line

Keep 1 or 4 cm (1/4 or 1½″) out-side 2–8 for hem or inturns

9–2 = 1.25 cm (1/2″). Join 9–8

10–8 = half of 8–9 plus 1.5 cm (1/2″)

11–10 = one-sixth chest, cuff opening = 5–6

Cuff

Length = one-fourth chest plus 1.5 cm (1/2″) or to taste

Width = one-sixteenth chest plus 1.5 cm (1/2″) or to taste

Single cuff

Square the lines 12 fold at 13–12

13–12 =cuff width plus 1.25 cm (1/2″) or nearly 7.5 cm (3″)

14–12 = half cuff length plus 1 cm (1/4″), or one-eighth chest plus 1 cm (1/4″)

15–13 = same as 14–12 or less 0.75 cm (1/4″)

Join 15–14

Shirt collar

Square lines from 0, fold at 0–2

1–0 = 6 cm (21/4″)

2–1 = 4 cm (11/2″)

3–0 = half neck plus 1 cm (1/4″)

4–3 = same as 1–0

5–4 = same as 2–1

6–4 = 4 cm (11/2″)

7–6 = same as 2–1

8–7 = 1 cm (1/4″)

9–5 = 5 cm (2″)

10–9 = 1 cm (1/4″), Shape 8–5–10–2 as shown

11–6 = 1.5 cm (1/2″)

12–11 = 0.75 cm (1/4″). Join 12–8

13–4 = 1 cm (1/4″)

Join 13–14 and produce the line to 15

15–13 = 10–13 cm (4–5″), shape 15–0 as shown

16–1 = 1 cm, Shape 16–13–12

Pocket

17–1 = about 5 cm (2″)

18–17 = one-eighth chest or plus 1.5 cm (½″), width

Pocket height = same as 18–17 plus 1.5 cm (½″)

18–19 and 17–20 = pocket height

Mark pocket as shown

Measurements details
Pant

Full length	:	35″
Inside leg	:	29″
Waist	:	28″
Seat	:	33″
Bottom	:	15″
Belt width	:	1½″

Method of drafting
Trouser

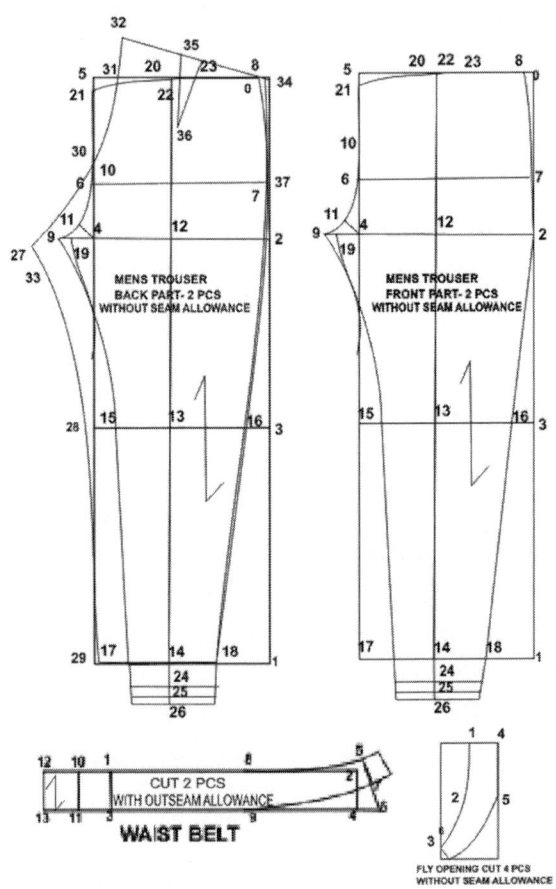

Front

Square lines from 0

1–0 = Full-length, less belt-width plus 1 cm (¼")

2–1 = inside leg plus 1 cm (¼")

3–2 = half inside leg less 5 cm (2")

Square lines from all these points

4–2 = one-fourth seat plus 4 cm (1½")

5–0 = same as 4–2. Join 4–5

6–4 = one-twelfth seat7–2 = same as 6–4. Join 6–7

8–5 = one-fourth waist plus 7.5 (3") for pleats and seams

Shape 8–7 as shown

9–4 =one-twelfth seat less 1.5 cm (½")

10–4= one-eight seat plus 1.5 cm (½")

11–4 half of 9–4 plus 0.75 cm (¼")
Shape fork 10–11–9
12 is mid-way 9–2
Square down from 12 to 13–14
17–14 = one- fourth bottom
18–14 = same as 17–14
19–9 = 2 cm (¾")
Join 19–17 and shape 9–15
16–13 = same as 15–13
Join 16–18 and shape 7–16
20 is squared up from 12
21–5 = 1 cm (¼"), shape 20–21
22–20 = 4 cm (1½") for pleat
23 is mid-way 22–8
Suppress 2.5 cm (1") in the pleat at 23
24–14 and 25–24 = each 4 cm 1 (1½")
26–25 = 2.5 cm (1")
Draw lines parallel to 17–18 from points 24, 25 and 26
Mark pock at a distance of 3 cm (1¼") below 8
Pocket opening = 1/6 seat

Back

27–9 = 5–6.5 cm (2–2½")
28–15 = 2.5 cm (1")
29–17 = 2.5 cm (1")
Shape 27–28 and join 28–29
30–10 = 2.5 cm (1") for seat angle. This quality should be increased for an erect figure and reduced for a stooping figure
Join 19–30 and produce to 31–32
32–31 = 2.5–4 cm (1–1½") according to flat or prominent seat
33–27 = 1 cm (1/4")
Shape fork 30–33 as shown
34–32 = one-fourth waist plus 4 cm (1½")
Draw side-seam 34–16–18
35–34 = one-twelfth seat plus 1.5 cm (½")
36 is squared down from 35 and equals 9 cm (3½")
Add 10–11.5 cm (4–4½")
Below 18–29

Waist belt

2–1 = Half waist plus 1 cm (1/4")
3–1 = 3.25 cm (1¼") or taste
4–2 = same as 3–1. Join 3–4

5–2 = 1.25 cm (½″)
6–4 = same as 5–3. Join 5–6
7–5 = same as 3–1 or less 1 cm (¼″)
8–2 and 9–4 = one-sixth waist
Shape 5–8 and 7–9
Draw 10–11 at a distance of 4 cm (1½″) from 1–3, the extension for the right side

If an extended belt is desired extend the left side to 12–13 which is 8–12.5 cm (3–5″) from 1–3

Keep about 4 cm (1½″) inlays outside 5–6, see cross-line

Fly-opening
1–4 = 5 cm (2″)
5–2 = 5 cm (2″)
6–3 = 1.5 (1/2″) Shape 4–5–6

Fabric required for construct this garment
Khaki (Brown) colour cotton plain fabric, lining material of same colour fabric, approximately 5.5 m fabric required for this garment.

Construction procedure
- Join the yoke part to back body part.
- Front left side finish with canvas for buttonhole side.
- Front right side finishes ¾″ for button fixing.
- Attach the pockets with flap at front left and right side.
- Join front and back pieces to each other on right and left side.
- Finishes the sleeves bottom hem.
- Joining the sleeves on armhole shapes.
- Match the front and back bodice and stitch the side seams.
- Complete bottom hem finish, attach a collar stand and collar with canvas finishing.
- Finishing back side dart, attach zippers front placket.
- Attach the side and back pockets.
- Finish the crotch point, joining the side seam and in leg seam. Finish the trouser bottom hem.
- Joint waist band to trouser part with canvas finishing.
- Fixing accessories and ironing.

Trims and accessories
Canvas, buttons, zipper, hook and eyelet.

Layout—Length-wise fold

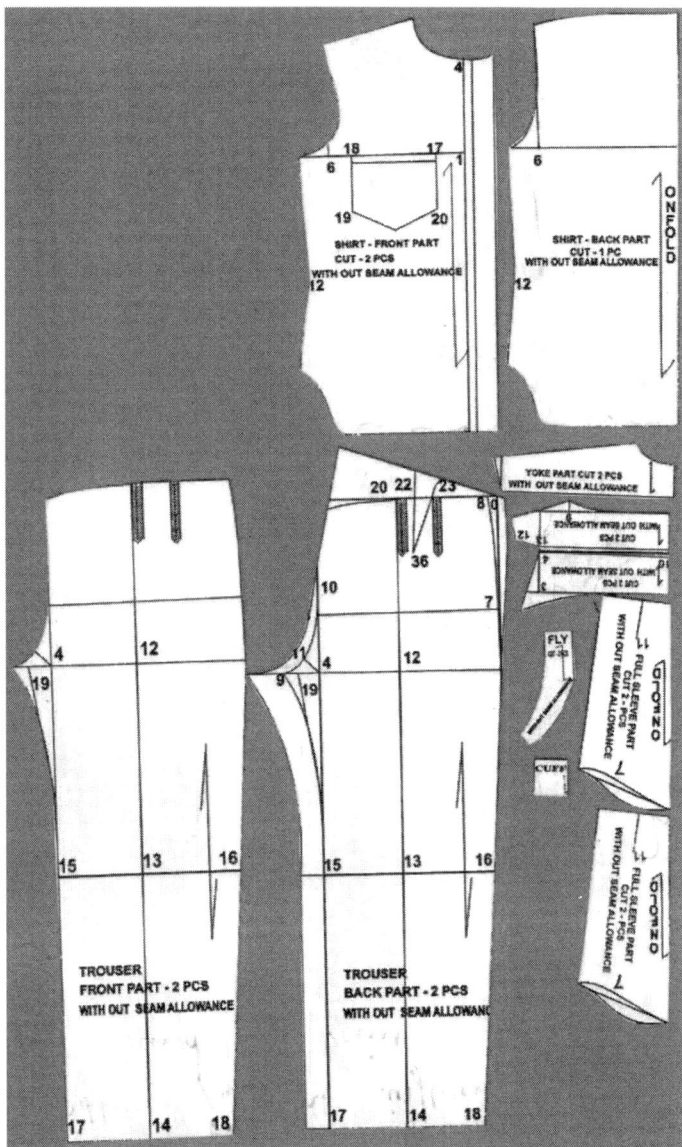

Cost calculation
Cost of material: Rs 600
Cost of accessories: Rs 120
Cost for construction: Rs 720
Total cost: Rs 1440

2.3   Design and construct uniform clothing for navy

A **shirt** is a cloth garment for the upper body (from the neck to the waist). The dress shirt is a button-up shirt with a stand collar and coat sleeves. The dress shirt is designed with pockets.

Trousers are worn on the lower part of the body from the waist to the ankles covering both legs separately. Peak cap, white tunic with shoulder straps for officers, captains and sailors. White tunic with arm badges, Gilt buttons size 1, Medals dress No 2, Ribbon dress No 4, Name tally, Chest badges, White trousers, White socks, White shoes.

Design of the garment

Aim

To design and construct the navy uniform.

Materials required

Fabric, pattern paper and tool kit.

Measurements details

Neck   :   35 cm (14″)
Chest   :   80 cm (32″)

Full length   :   76 cm (30″)
Shoulder   :   20 cm (8″)
Half sleeve   :   30 cm (12″)

Method of drafting
Navy shirt

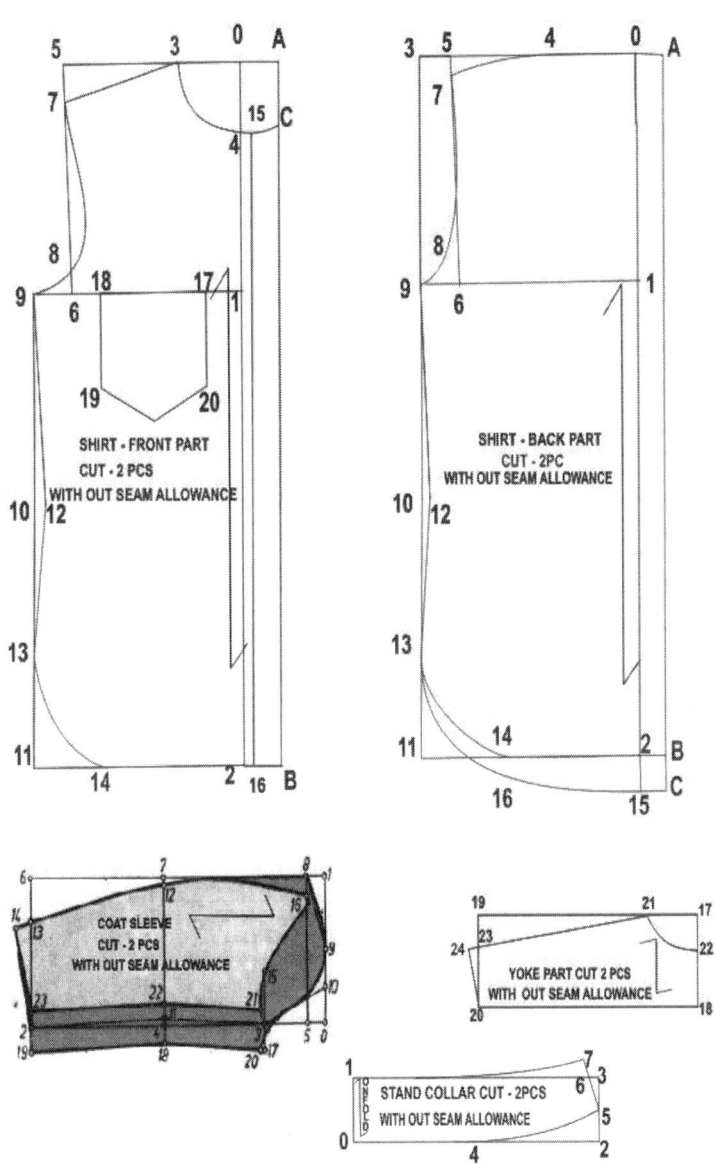

Front - 0A and 2 to B = 2.5″

Draw a line 0–1–2 at distance of 4 cm (1½″) from the fold A–B

1–0 = one-fourth chest

2–0 = full length plus 2.5 cm (1″) Square out from 0, 1 and 2

3–0 = one-sixth neck

4–0 = one-sixth neck plus 0.75 cm (1/4″) Shape neck 4–3 as shown

5–0 = shoulder plus 0.75 cm (1/4″)

6–1 = one-fourth chest join 5–6

7–5 = one-fourth of 5–6 join shoulder 3–7

8–6 = about 3 cm (1¼″)

9–1 = one-fourth chest plus 5 cm (2″) or taste

Shape scye 7–8–9

Square down from 9 to 10 and 11

10–9 = about one-fourth chest

12–10 = 1.5 cm (1½″)

13 is mid-way 10–11

14–11 = about one-eighth chest

Shape 9–12–13–14 as shown

15–4 = 1.5 cm (1½″)

16–15 (parallel to 1–0) = one-fourth chest plus 10 cm (4″)

Cut the lower layer from 15 to16 of the right side, i.e., the left side should be wider than the right side.

C–A = same as 3–0. Join 4–C.

Back

Cut the front and use it for back in such a way that lines 1–2 and 1–9 of front and back will be coincide. The back is cut without pleats or gatherings.

Square lines from 0, fold at 0–2.

1–0 = same as 1–0 of front less 4 cm (1½″)

2–1 = same as 2–1 of front, Square out from 0, 1 and 2

6–0 = same as 5–0 of front. Square down from 5–6

7–5 = 1.5 cm (½″)

4 is mid-way 5–0, Shape 7–4 as shown

Except 8, the proportions of 9–1 are the same like front

Shape scye 7–8–9

Stand collar

Square from 0, fold at 0–1

1–0 = 4 cm (1 ½″), or to taste

2–0 = half neckline plus 1 cm (1/4″)

3–2 = same as 1–0. Join 3–1

4 is mid-way 2–05–2 = 1.

5 cm (1/2″). Shape 5–4
6–3 = 1 cm (1/4″)
Join 5–6 and extended to 7
7–5 = 1–0 less 0.75 cm (1/4″)
Shape 7–1 as shown in drafting

Coat sleeve
Sleeve—Top side

Square lines from 0
1–0 = one-fourth chest less 1.5 cm (1/2″)
2–0 = sleeve length plus 2.5 cm (1″)
3–0 = one-eight chest plus 1.5 cm (1/2″)
4–3 = half of 2–3 less 2.5 cm (1″)
5–0 = 4 cm (1 ½ ″) Square out from 2, 4, 3 and 5
6–2 = same as 1–0. Join 1–6
9 is mid-way 1–10
10 is mid-way 9–0
Join 3–10
Shape 3–9–8 as shown
11–4 = 1.5 cm (1/2″)
Join 3–11 and 11–12
12–11 = one-sixth chest plus 4 cm (1½″)
13–2 = one-sixth chest
Shape 8–12–13 and produce to 14
14–13 = 3 cm (1 ¼″)
Join 2–14

Sleeve—Under side

15–3 = one-twelfth chest
16–8 = 4 cm (1½″)
Take 1 cm (1/4″) mid-way below 3–15 and shape 3–15–16 as shown
Shape 16–12
17–3, 18–11 and 19–2 = 3 cm (1¼″) each. The width to be increased for the top side
20–17 = 1 cm (1/4″)
Join 20–18–19 and shape 20–9–8
21–3, 22–11 and 23–2 = 2 cm (3/4″) each
Join 21–22, 22–23 and 23–14
Top side = 20–9–8–12–14–2–19–18–20
Underside = 21–15–16–12–14–23–22–21

Measurements detailsPant

Full length :   104 cm (41″)

Inside leg	:	73 cm (29″)
Waist	:	76 cm (30″)
Seat	:	92 cm (36″)
Bottom	:	43 cm (17″)
Belt width	:	4 cm (1½″)

Method of drafting
Trouser

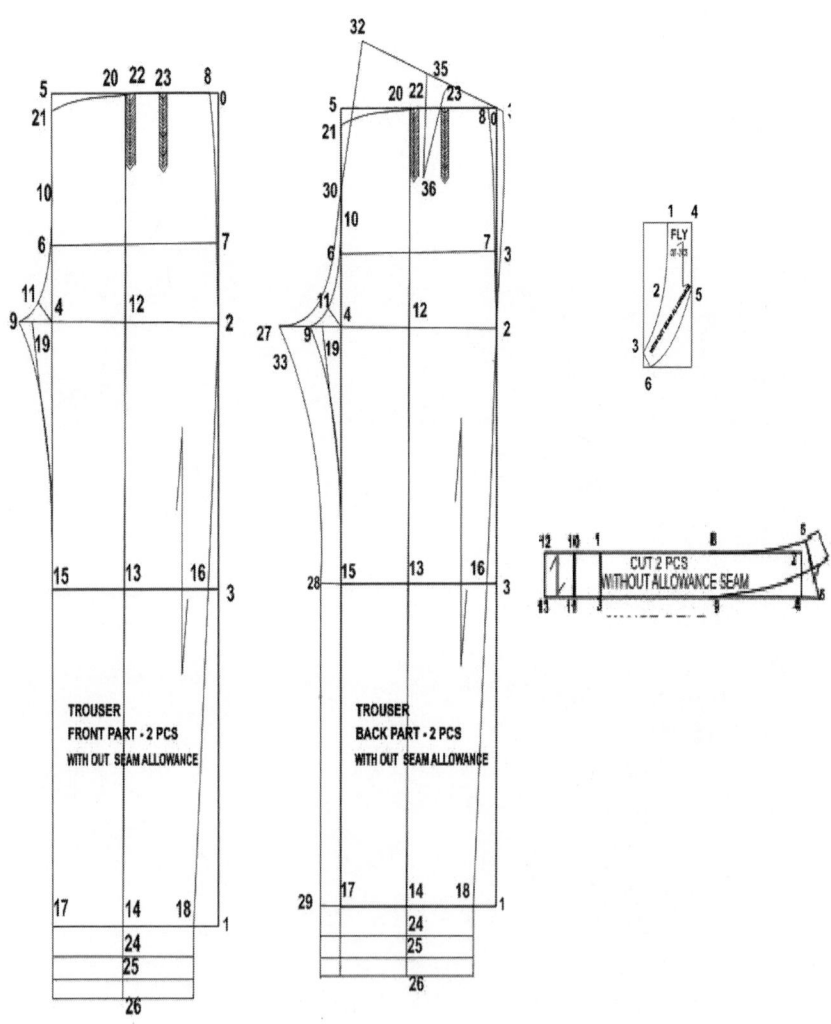

Front

Square lines from 0

1–0= Full-length, less belt-width plus 1 cm (¼")

2–1 = inside leg plus 1 cm (¼")

3–2 = half inside leg less 5 cm (2")

Square lines from all these points

4–2 = one-fourth seat plus 4 cm (1½")

5–0 = same as 4–2. Join 4–5

6–4 = one-twelfth seat

7–2 = same as 6–4. Join 6–7

8–5 =one-fourth waist plus 7.5 (3") for pleats and seams

Shape 8–7 as shown

9–4 = one-twelfth seat less 1.5 cm (½")

10–4 = one-eight seat pulse 1.5 cm (½")

Strike a line mid-way across the angle 10–4–911–4 = half of 9–4 plus 0.75 cm (¼")

Shape fork 10–11–9

12 is mid-way 9–2

Square down from 12 to 13–14

17–14 = one-fourth bottom

18-14 same as 17 to 14

19–9 = 2 cm (¾")

Join 19–17 and shape 9–15

16–13 = same as 15–13

Join 16–18 and shape 7–16

20 is squared up from 12

21–5 = 1 cm (¼") shape 20–21

22–20 = 4 cm (1½") for pleat

23 is mid-way 22–8

Suppress 2.5 cm (1") in the pleat at 23

24–14 and 25–24 = each 4 cm 1 (1½")

26–25 = 2.5 cm (1")

Draw lines parallel to 17–18 from points 24, 25 and 26

Mark pock at a distance of 3 cm (1¼") below 8

Pocket opening = 1/6 seat

Back

27–9 = 5–6.5 cm (2–2½")

28–15 = 2.5 cm (1")

29–17 = 2.5 cm (1")

Shape 27–28 and join 28–29

30–10 = 2.5 cm (1") for seat angle

This quality should be increased for an erect figure and reduced for a stooping figure

Join 19–30 and produce to 31–32

32–31 = 2.5–4 cm (1–1½") according to flat or prominent seat

33–27 = 1 cm (1/4") shape fork 30–33 as shown

34–32 = one-fourth waist plus 4 cm (1½")

Draw side-seam 34–16–18

35–34 = one-twelfth seat plus 1.5 cm (½")

36 is squared down from 35 and equals 9 cm (31½")

Add 10–11.5 cm (4–4½")

Below 18–29

Waist belt

2–1 = Half waist plus 1 cm (1/4")

3–1 = 3.25 cm (1¼") or taste

4–2 = same as 3–1. Join 3–4

5–2 = 1.25 cm (½")

6–4 = same as 5–2. Join 5–6

7–5 = same as 3–1 or less 1 cm (¼")

8–2 and 9–4 = one-sixth waist

Shape 5–8 and 7–9

Draw 10–11 at a distance of 4 cm (1½") from 1–3, the extension for the right side

If an extended belt is desired extend the left side to 12–13 which is 8–12.5 cm (3–5") from 1–3

Keep about 4 cm (1½") inlays outside 5–6, see cross-line

Fly-opening

1–4 = 5 cm (2")

5–2 = 5 cm (2")

6–3 = 1.5 (1/2"), Shape 4–5–6

Fabric required for construct this garment

White colour cotton plain fabric, approximately 5 m fabric required for this garment.

Layout—Lengthwise-fold

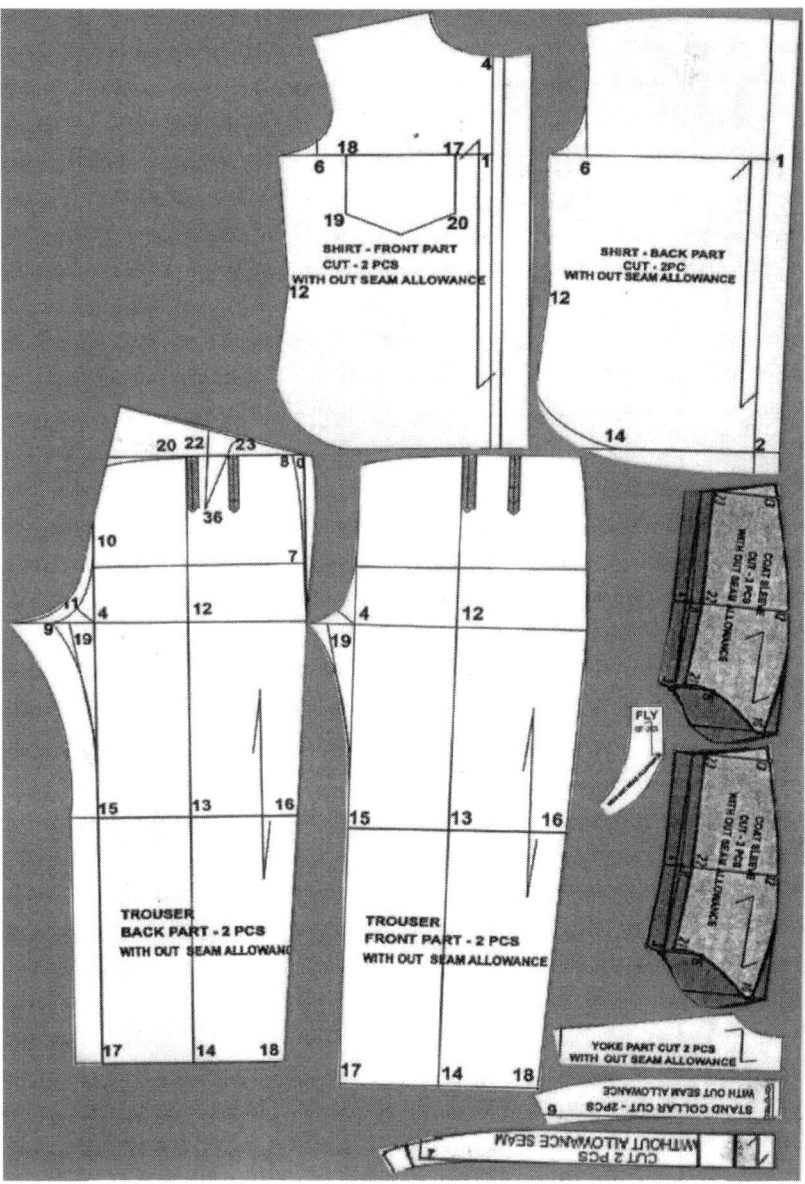

Construction procedure
- Join the yoke part to back body part.
- Front left side finish with canvas for buttonhole side.
- Front right side finishes ¾″ for button fixing.
- Attach the pockets with flap at front left and right side.
- Join front and back pieces to each other on right and left side.
- Finishes the sleeves bottom hem. Joining the sleeves on armhole shapes.
- Match the front and back bodice and stitch the side seams.
- Complete bottom hem finish, attach a collar stand and collar with canvas finishing.
- Finishing back side dart, attach zippers front placket.
- Attach the side and back pockets.
- Finish the crotch point, joining the side seam and in leg seam.
- Finish the trouser bottom hem.
- Joint waist band to trouser part with canvas finishing.
- Fixing accessories and ironing.

Trims and accessories
Canvas, buttons, zipper, hook and eyelet

Cost calculation
Cost of material: Rs 850
Cost of accessories: Rs 180
Cost for construction: Rs 800
Total cost: Rs 1830

2.4   Design and construct uniform clothing for doctor

A white coat or lab coat (often popular as apron) is a knee-length overcoat worn by professionals in the medical field or by those involved in laboratory work. The garment is made from white or light-coloured cotton, linen, or cotton polyester blend, allowing it to be washed at high temperature and make it easy to see if it is clean. The coat protects their street clothes and also serves as a simple uniform.

Design of the garment

Aim

To design and construct the doctor coat.

Materials required

Fabric, pattern paper and tool kit

Measurements details

Chest	:	92 cm (36″)
Waist	:	82 cm (32″)
Seat	:	96 cm (38″)
Full length	:	76 cm (30″)
Waist length	:	41 cm (16″)
Half back	:	19 cm (7½″)
Shoulder	:	21.5 cm (8½″)
Sleeve with shoulder	:	80 cm (31″)

Method of drafting

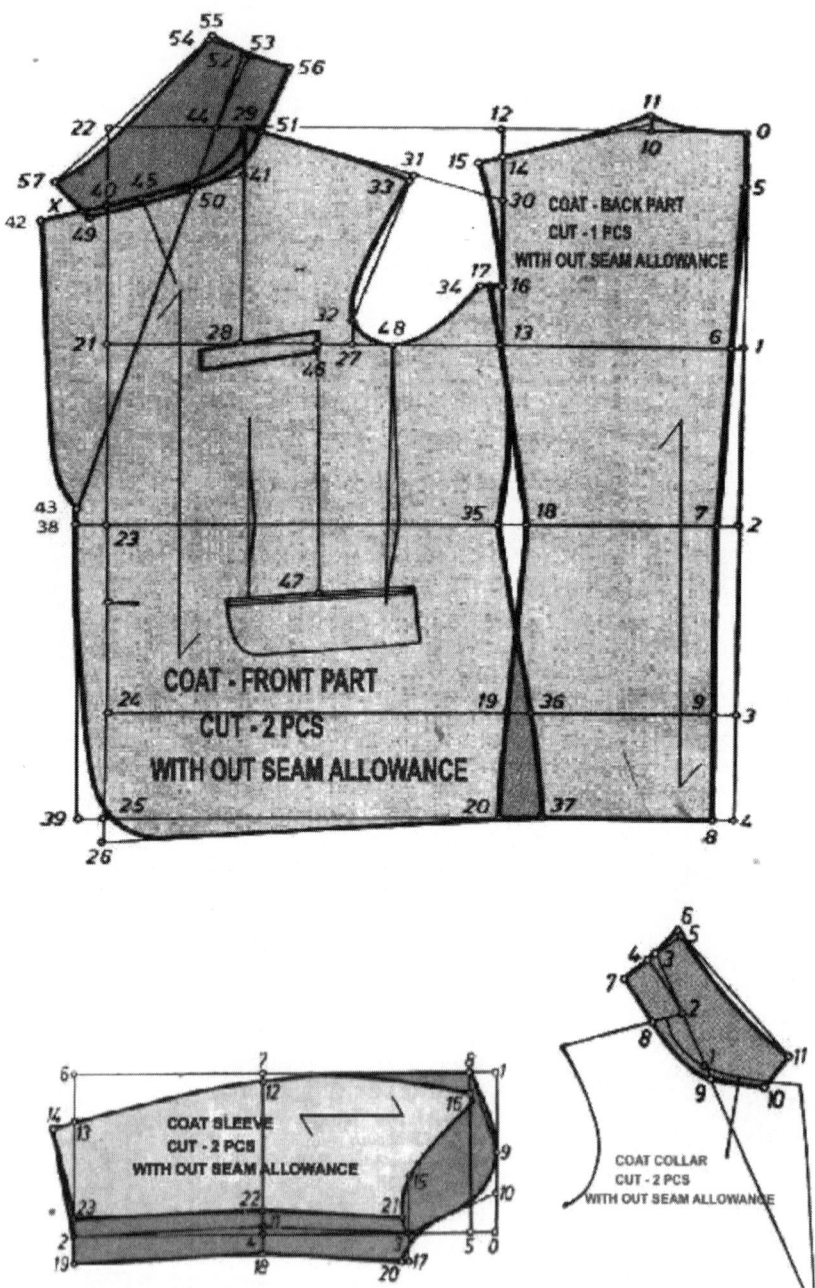

COAT - BACK PART
CUT - 1 PCS
WITH OUT SEAM ALLOWANCE

COAT - FRONT PART
CUT - 2 PCS
WITH OUT SEAM ALLOWANCE

COAT SLEEVE
CUT - 2 PCS
WITH OUT SEAM ALLOWANCE

COAT COLLAR
CUT - 2 PCS
WITH OUT SEAM ALLOWANCE

Back part

    Square lines from 0

    1–0 = one-fourth chest

    2–0 = waist length

    3–2= Half waist length

    4–0 = Full length, square out from 1, 2, 3 and 4

    5–0 = one-sixteenth chest or one-fourth of 1–0

    7–2 and 8–4 = 2 cm (¾″) each

    Join 5–7 and 7–8

    10–0 = one-twelfth chest

    11–10 = 2 cm (¾″)

    Shape neck 0–11

    12–0 = half back plus 1.5 cm (½″)

    Square down from 12–13

    14–12 = 3 cm (1¼″)

    Join 11–14 and produce to 15

    15–14 = 2 cm (¾″)

    16–13 = 6.5 cm (2½″), Shape 15–16

    17–16 = 1 cm (¼″)

    18–7 = one-sixth chest

    19–9 = 18–7 plus 2 cm (¾″)

    20–8=7″

    Shape side seam 17–18–19–20

Front part

    21–6 = Half chest plus 6.5 cm (2½″)

    22–25 is squared from 21

    26–25 = 2.5 cm (1″)

    27–21 = one-fourth chest less 2.5 cm (1″)

    28–27 = one-twelfth chest plus 1.5 cm (½″), Square up from 28–29

    30–12 = one-sixteenth chest plus 2 cm (¾″), Join 29–30

    31–29 = 11–15 less 1 cm (¼″)

    32–27 = 2.5 cm (1″)

    Join 31–32. 33–31 = 1 cm (1/4″)

    Shape shoulder 29–33

    34–17 = 1 cm (¼″), Shape scye 33–32–48–34

    35–23 = half waist plus 7.5 cm (3″) less 7–18 of back

    36–24 = half seat plus 5 cm (2″) less 9–19 of back

    Shape side seam 17–13–35–36–37

    38–23 and 39–25 = 2.5 cm (1″)

    Each join 38–39

    40–22 = one-twelfth chest

41–29 = one-twelfth chest less 2.5 cm (1″)

Join 41–40 and produce to 42

42–40 = 6 cm (2¼″) or to taste

42–38 = 1.5 cm (½″)

44–29 = 2.5 cm (1″). Join 43–44

45–40 = 3 cm (1¼″)

Take a small dart at 45 of 7.5–9 cm (3–3½″) long

46–27 = 2.5 cm (1″)

Upper pocket in front of 46 = 10 cm (4″) long and 2 cm (¾″) wide

Squarer down from 46–47

47 from the waistline = one-twelfth chest

47 is the mid-point of lower pocket

Pocket opening is about 15 cm (6″)

48–27 = 4 cm (1½″)

Front dart

Mark the dart 2 cm (¾″) inside the front edge of the lower pocket and parallel to line 22–25

The upper point of this dart is about 6.5 cm (2½″) below the welt pocket. Suppress about 1 cm (3/8″) in this dart.

Underarm dart

Mark the dart 2 cm (3/4″) inside the rear edge of the lower pocket from 48. Suppress 1.5–2 cm(½″–¾″) in this dart.

Coat collar

Using the neck curve of the front, dart the collar as follows:

X - 42 = 4 cm (1½″)

49 - X = 1 cm (¼″), point 50 is 1 cm (¼″) below the neck curve and on the crease line.

51–29 = about 1.5 cm (½″)

Produce line 43–44–52

52–44 = one-twelfth chest plus 1 cm (¼″)

53–52 = 1 cm (¼″), shape 53–44

54–56 is square from 53–44

54–53 = 4–5 cm (1½″–2″)

55–54 = 1 cm (¼″), Join 55–53

56–53 = 3–4 cm (1¼–1½″), Shape 49–50–51–56

57–X = 4 cm (1½″), Join 57–55 and shape as shown

The straight grain of canvas should come at 1–2 as shown in the figure.

Sleeve—top side

Square line from 0.

1–0 = one-fourth chest less 1.5 cm (½″)

2–0 = Sleeve length plus 2.5 cm (1″)
3–0 = one-eighth chest plus 1.5 cm (½″)
4–3 = Half of 2–3 less 2.5 cm (1″)
5–0 = 4 cm (1½″), Square out from 2, 4, 3 and 5
6–2 = same as 1–0. Join 1–6
9 is mid-way 1–0
10 is mid-way 9–0. Join 3–10, Shape 3–9–8 as shown
11–4 = 1.5 cm (½″), Join 3–11 and 11–2
12–11 = one-sixth chests plus 4 cm (1 ½″)
13–2 = one-sixth chest, shape 8–12–13 and produce to 14
14–13 = 3 cm (1¼″), Join 2–14

Sleeve under

15–3 = one-twelfth chest
16–8 = 4 cm (1½″)
Take 1 cm (¼″) mid-way below 3–15 and shape 3–15–16 as shown.
Shape 16–12
17–3, 18–11 and 19–2 = 3 cm (1¼″) each, the width to be increased for
the top side
20–17 = 1 cm (1/4″)
Join 20–18–19 and shape 20–9–8
21–3, 22–11 and 23–2 cm (¾″) each
Join 21–22, 22–23 and 23–14
Top side = 20–9–8–12–14–2–19–18–20
Underside = 21–15–16–12–14–23–22–21

Collar

Trace the outline of the front neckline and label as shown above and draft the
collar as follows:

In the above draft, point 1 is taken on a label fold-line at the neckline and
point 2 is taken where shoulder and label line meets when extended.
Join 1–2 and extend up to 3
3–2 = back neckline (i.e., 11–0 in S.B. Coat draft) plus 1 cm (¼″)
4–3 = 1.5 cm (½″), Join 4–2
Line 5–4–7 is square from 4–2
5–4 = 4 cm (1½″)
6–5 = 1.5 cm (½″), shape 6–4
7–4 = 3 cm (1¼″)
Point 8 is taken 1.5 cm (½″) inside neck curve
Point 9 and 10 are taken 0.75 cm (¼″) inside neck curve
Shape 7–8–9–10

Fabric required for construct this garment

White colour cotton plain fabric approximately 2.5 m fabric required for this garment.

Layout—Lengthwise layout

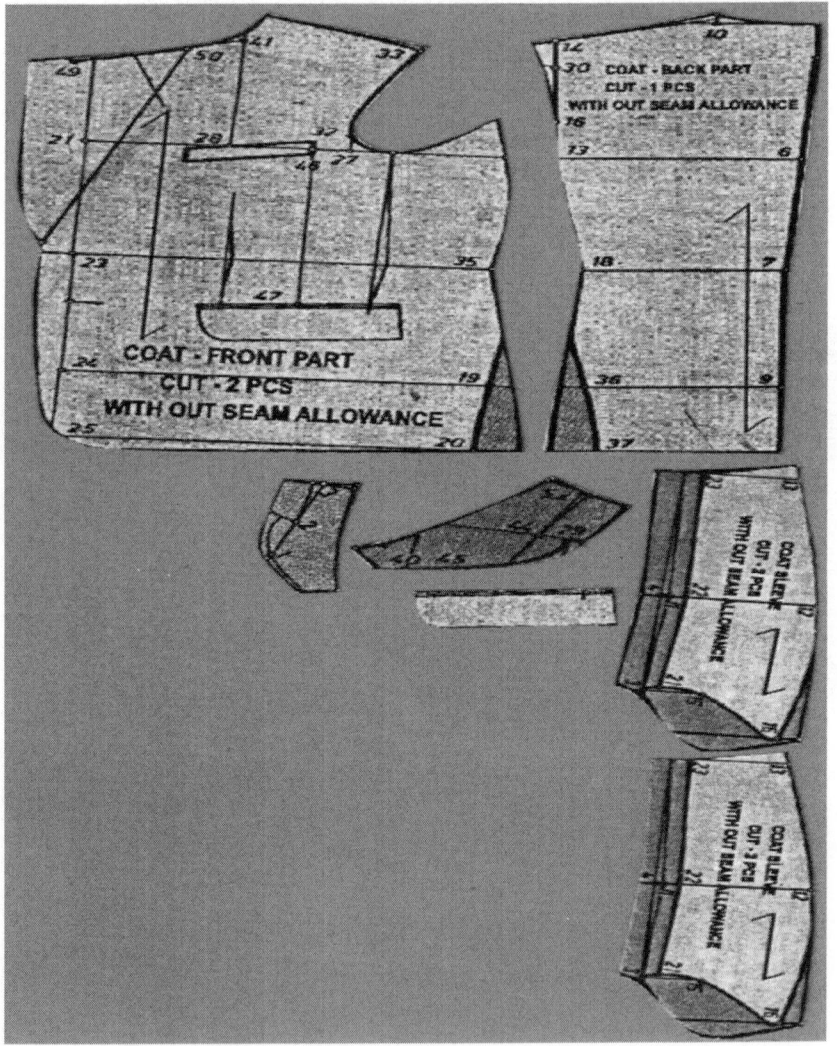

Construction procedure

- Finishing armhole and front dart.
- Attach the pockets at front left and right side.
- Front left side for button hole side.

- Front right side finishes ¾" for button fixing.
- Join front and back pieces to each other on right and left side.
- Finishes the sleeves bottom hem.
- Joining the sleeves on armhole shapes.
- Match the front and back bodice and stitch the side seams.
- Complete bottom hem finish.
- Attach a collar stand and collar with canvas finishing.
- Fixing accessories and ironing.

Trims and accessories
Canvas, buttons.

Cost calculation
Cost of material: Rs 300
Cost of accessories: Rs 60
Cost for construction: Rs 250
Total cost: Rs 610

2.5   References

1. Zarapkar Tailoring College. Zarapkar System of Cutting. Navneet Publications (India) Limited, Dantali, Gujarat. Printed by Shreeji offset, 99, Amrut Industrial Estate, Ahmadabad, 380052. ISBN 81-243-0199-9. pp 117, 124–126, 149, 152, 154–156, 175–178.2.

2. Anita Tyagi. Garment Construction. Sonali Publications, New Delhi-110002, ISBN 978-81-8411-487-4.

PART II

Fashion show garments

3.1 Introduction

The word fashion to the Latin "facere" which means "to make" or "to do" in clothing. Fashion is a style that is popular in the present time. The Webster International Encyclopaedia in 1998 defines fashion as "prevailing style of dress, new designs representing changes from previous seasons". Fashion is by its very nature, an ever-changing art, Oscar Wilde remarked that "Fashion is a form of ugliness so intolerable that we have to alter it every six months" but it is this continual evolution, the constant reinvention of old trends and the creation of new ones, that lends the fashion industry its excitement and glamour. A fashion show is an event put on by a fashion designer to showcase their innovative clothing. An innovative collection is produced by a designer, brand, company, group of people, etc. The fashion parade of moving bodies has an essential feature of a fashion show and has given rise to the modelling profession. Fashion shows represent every season such as the summer, spring, winter and autumn.

The fashion show should have a **theme**, a central basic idea on which the show is built. The theme will say about the type of merchandise being shown and the audience from whom the show is prepared. A theme suggest signs, publicity and programme. Fashion show should be a theme, a logo, a concept or a symbol that is carried through the whole show from the preliminary announcements until the final. In fashion show light should follow the model. If this is not possible, the stage or runaway should be in light. If fashion show runs longer than fifteen minutes, it is desirable to have variation in lighting.

Fashion show is a chance for a designer to show-off a particular aesthetic, a particular mood, a particular feel (or) point of view. As a result, fashion shows can tend to be more conceptual and focused on a high level idea.

3.2 Design and construct for fashion show selected five themes

1. Barbie
2. Rainbow
3. Peacock
4. Cheetah
5. Cherry blossom

Barbie

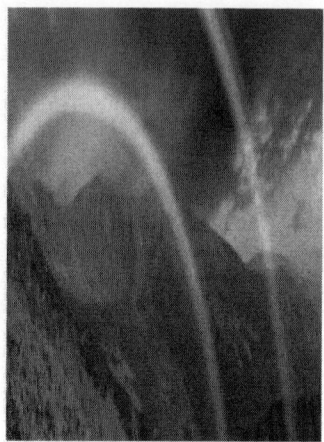

Rainbow

Peacock

Cheetah

Cherry blossom

3.3   About fashion portfolio

Portfolio should include your best work, from classes, internships and any professional projects. It should be organised, consistent and demonstrate your skills. Start with the beginning inspiration of a project, to the final outcome, (do not show processes). All work should be finished. For the fashion designer seeking employment, a well-prepared portfolio is an essential marketing tool. Designing Your Fashion Portfolio: From Concept to Presentation uses the design process to guide students through conceptualisation and assembly of a fashion design portfolio that will communicate their talents and vision as designers.

The richly illustrated text helps students assemble their work and organise it into a compelling story of their artistic talents. In the process, students learn to evaluate their skills and identify their interests so that they can focus on building collections for their chosen target markets. The author's fashion design portfolio system enables designers to tailor their portfolios for each client throughout their careers.

3.4   Introduction

A collection of recent work, made-up of different media including illustration, photography, design, print, sketch books and more, showing your development as you progress through your studies as an artist.

Design development process are as follows:

- Research
- Theme/Mood board
- Inspiration board
- Illustration board
- Client board
- Accessory board
- Colour board
- Flat sketches board
- Garment construction pattern details
- Trim and fabric board
- Muslin fit
- Costing
- Presentation
- Photos
- Conclusion

3.4.1   Research

Research is a process of seeking and recording creative information in order to compile a bank of visual information for inspirational purposes. The process of recording research material provides a focus from which to start generating ideas, which can come from anywhere, they can be completely original and be only connected to the designer in a very individualistic way.

Research can be divided in to two types: investigative research which involves seeking and recording information from a wide range of reference points, for example, historical sources, museums, exhibitions, shops and collecting materials, photographing details of construction techniques, exploring a specific area in depth. The second is inspirational, this can be drawn or photographed from any source and is often a wide ranging selection of images, materials, colour schemes, articles, sketches, fabrics, notes, scraps of wrapping paper,wallpaper,advertisements,photographs,trimmings,articles,sewn samples, memorabilia, postcards, old patterns, video, animation clips, music and graphics, anything that is aesthetically and thematically inspiring.

Research should add to your understanding of the market and customer, showing employers that you understand and care about this essential aspect of design. Before starting your work, research about the latest fashion trends, colour and work backwards through your fashion career history.

3.4.2   Theme/Mood board

Mood boards are usually a collection of images, texture, colour, etc., compiled with the intention of communicating a visual statement provoking a mood to inspired designs. A mood board is a summary of your collections inspiration and theme. It's a design tool that will help you remain focused and consistent as your line develops. It's also a great communication aid when explaining your vision to others. Magazine tears, fabric swatches, old photos, buttons, ribbons, basically any visual reference you desire are mounted on to a hard board. Be sure to give your story board a title, like a book or film.

3.4.3   Inspiration board

Fashion requires the "**fresh**" to inspire design and offer something new to the customer. A key element of any new season is what inspires that season. In a visually rich world there is a constant need to keep stimulating the fashion consumer be it through mood, print, pattern, texture. Gathers all the reference points for what the product will look like. More specific and visual collection of visual references that are the starting point for elements. Inspiration board should respond to the mood board.

3.4.4   Illustration board

Show your sketches separately. This will assist those of us who think of your sketching process as one of the most important and telling parts of your presentation. The illustrated poses deliberately have attitude relevant to the theme. The look is quite youthful, casual, utilitarian and urban in outlook.

3.4.5   Client board

Gives an idea of your client, his or her likes and dislikes. The customer can be an individual or a particular group, you also have to keep in mind, the age of the client, the educational and the economic background of your client. At the same time do not be too specific as in naming your client, giving his address, etc.

3.4.6   Accessory board

Rapidly changing sector focuses on bags, belts, leather goods, jewellery, eye wear and watches. The accessories can be categorised mainly into three types as follows:

Carried accessories
Handbags, hand fans, swords, parasols, umbrella, etc.

Worn accessories
Jackets, boots and shoes, ravats, ties, sunglasses, belts, shawls, scarves, stocks, stockings, gloves, muffs, jewellery, watches, etc.

Detachable accessories
Aigrettes, lapels, pins and bags, etc.

3.4.7   Colour board

As colour is the first development of a new season, colour stories are often expressed using whatever colour samples are available, for example, flat fabrics, yarn samples, swatch samples, etc., colour palettes are inspired by exhibitions, gallery visits, interest in particular cultures, unusual imagery or historical imagery.

3.4.8   Flat sketches board

Fashion illustrations are a fun way to explain the attitude of a garment and show-off your artistic talents. These images should show your individual "**hand**". Illustrations should also showcase the most important feature of the garment. Illustrations are a way to show that you understand how certain fabrics drape, pleat or hold form. This is also a great way to show you understand how clothes fit the body. Black and white drawings of **"flats"**explain more about the detail and shape of the story's garments.

3.4.9   Garment construction pattern details

The pattern details for the designed garment is highlighted in this board.

3.4.10   Trim and fabric board

Trim and fabric board consist of embellishments, closures, laces, threads, fusing material, label and fabric swatches.

3.4.11   Muslin fit

Muslin fit are done by either using draping method or pattern making methods. This is the process of test fit done before actualising on real fabric. It helps in correcting any ill fit or flaws in the garment. It done on cheaper fabrics. Muslin fit or toile are transferred on to real fabrics after alteration or correction if any. The transferring is done with the help of tracing wheel, scales and markets.

3.4.12   Costing

A cost sheet is a complete record of each design and is used to cost the garment and establish the wholesale price. Includes all expenses inherited during the development of every single garment piece. Later a price is put up on the basis such that all the expenses + profit is incurred back can hike any percentage of the total net expense to gain back any percentage of profit.

3.4.13   Presentation

Collection launch at a fashion runway. Retail store exhibition. Exhibition to client. E-store launch. Presenting to the in-house head designer. With increasing demand for **fast fashion** companies convert the latest runway trends.

3.4.14   Photos

Photos are optional supplement to the portfolio, but they are always a **big plus**. By having them you will show the jury or employer that you not only have the ideas, but also the skills to bring your designs to life. If you have produced the clothes it is very important to make professional looking photos.

3.4.15   Conclusion

Fashion keeps inviting and exploring new lands and base when every time a new design develops. But the concepts and process involved remains the same and design derivation are always fresh, crisp and edgy. Fashion will never fade its face in any era. Fashion in you showcase the outlook who you are as a person.

3.5   Barbie theme

Research—Barbie

Theme board/Mood board

Inspiration board

Illustration board

Client board
Age group: 20
Occasion: Party wear
Profession: Artist
Income per month: Rs30,000
Market: National
Description: It is a barbie typed outfit. Unique outlook.

Accessory board

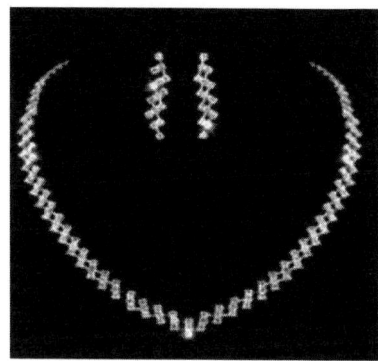

Colour board

Flat sketches board

Garment construction pattern details

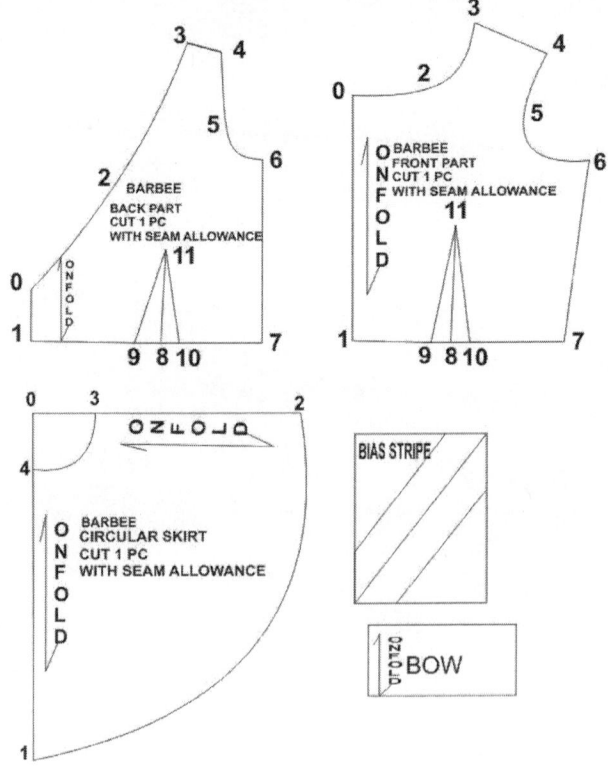

Trim and fabric board

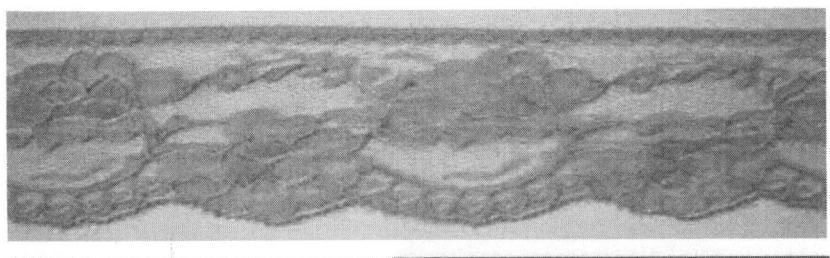

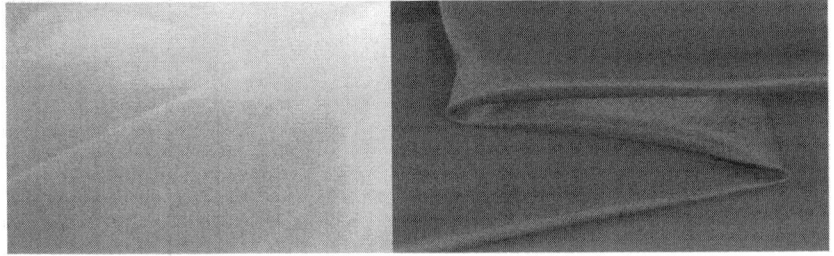

Muslin fit

Sewing a muslin is a great way to test out the construction and fit of a sewing pattern without the risk of wasting good fabric and a lot of time on a garment.

Costing

Includes all expenses inherited during the development of every single garment piece.

Material cost

Materials in meters	Price	Total amount
Net material–15	90	1350
Satin material–15	120	1800
Lining material–10	70	700

Trims and accessories

Trims	Price	Total amount
Ear rings	250	250
Decorative chain	700	700
Slipper	1500	1500
Bracelet	450	450

Labour cost for stitching:Rs1000
Total cost of the garment: Rs7750

Presentation
Collection launch in a fashion runway.

Photos

Conclusion
A toy that presents a sexiest play boy image of women but a toy that is independent and more liberated than traditional baby dolls. The barbie colour garment trapped in barbies world with good ending in portfolio.

3.6   Rainbow theme

Research—Rainbow

Theme board/Mood board

Inspiration board

Illustration board

Client board
Age group: 18
Occasion: Party wear
Profession: Artist
Income per month: Rs25,000
Market: National
Description: Colourful layers with red top filled with ruffles and pleats. Gorgeous and elegant outlook.

Accessory board

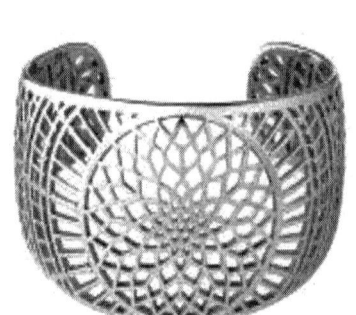

Colour board

Flat sketches board

Garment construction pattern details

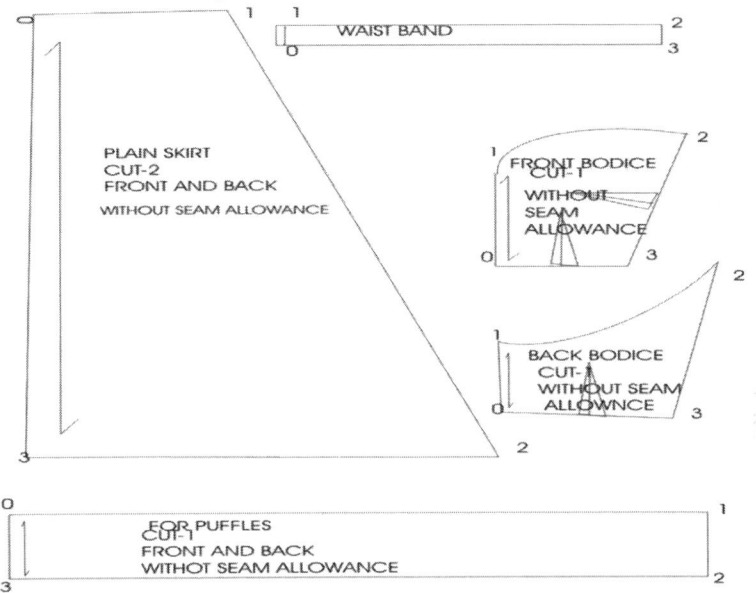

Trim and fabric board

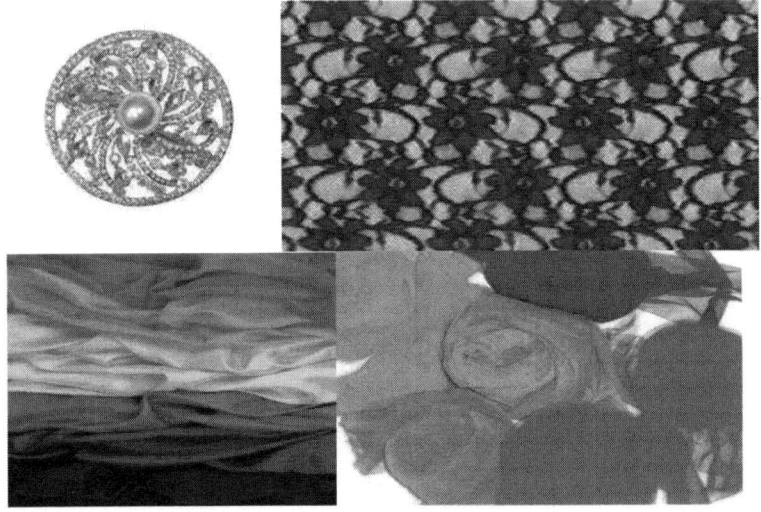

Muslin fit

Sewing a muslin is a great way to test out the construction and fit of a sewing pattern without the risk of wasting good fabric and a lot of time on a garment.

Costing
Includes all expenses inherited during the development of every single garment piece.

Materials in meters	Price	Total amount
Net material–15	90	1350
Satin material–3.5	120	420
Lining material–4	45	180

Trims and accessories

Trims	Price	Total amount
Crowns	250	250
Decorative chain	300	300
Bangles and stead	400	400
Slipper	700	700

Labour cost for stitching:Rs1000
Total cost of the garment: Rs4600

Presentation
Collection launch in a fashion runway.

Photos

Conclusion

Thus fashion show garment created with the rainbow theme presentations are always a highlight.

3.7   Peacock theme

Research—Peacock

Theme board/Mood board

Inspiration board

Illustration board (without sleeve)

Client board
Age group: 17
Occasion: Party wear
Profession: Student
Market: National
Description: Inspired by peacock paintings. Fabric paints are used with multiple decorative stones.

Accessory board

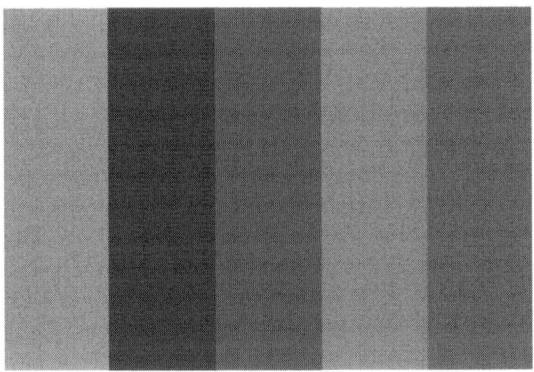

Colour board

Flat sketches board

Garment construction pattern details

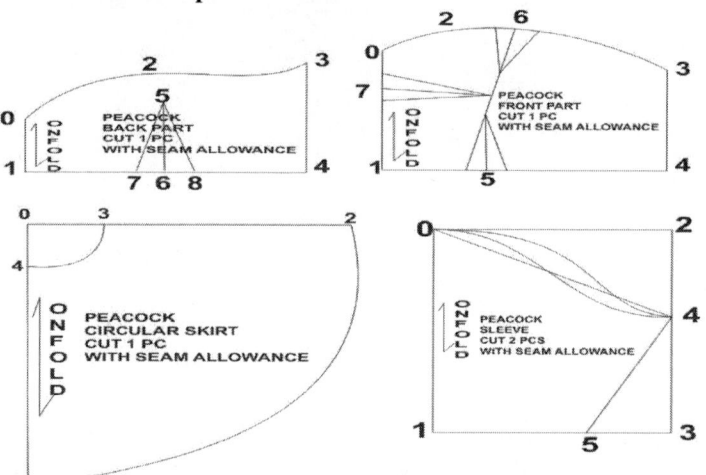

Trim and fabric board

Muslin fit

Sewing a muslin is a great way to test out the construction and fit of a sewing pattern without the risk of wasting good fabric and a lot of time on a garment.

Costing

Materials in meters	Price	Total amount
Satin–7	120	840
Lining material–5	70	350

Trims and accessories

Trims	Price	Total amount
Bangle	250	250
Slipper	1200	1200
Chain	150	180
Earrings	200	200

Labour cost for stitching—Rs 1000
Total cost of the garment—Rs 4000

Presentation

Collection launch at a fashion runway.

Photos

Conclusion
A new design of peacock-inspired life in contemporary trends.

3.8   Cherry blossom theme

Research—Cherry blossom

Theme board/mood board

Insipiration board

Illustration board

Client board
Age group: 19
Occasion: Party wear
Profession: Student
Market: National
Description: Inspired by rose flower and cherry colour.

Accessory board

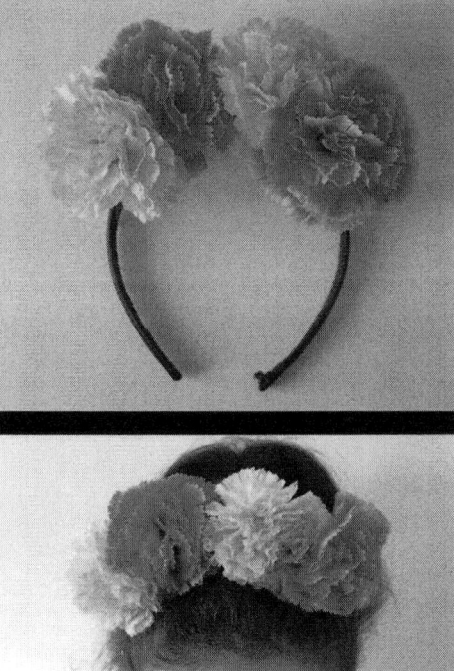

Colour board

Flat sketches board

Garment construction pattern details

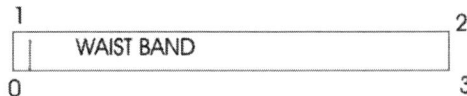

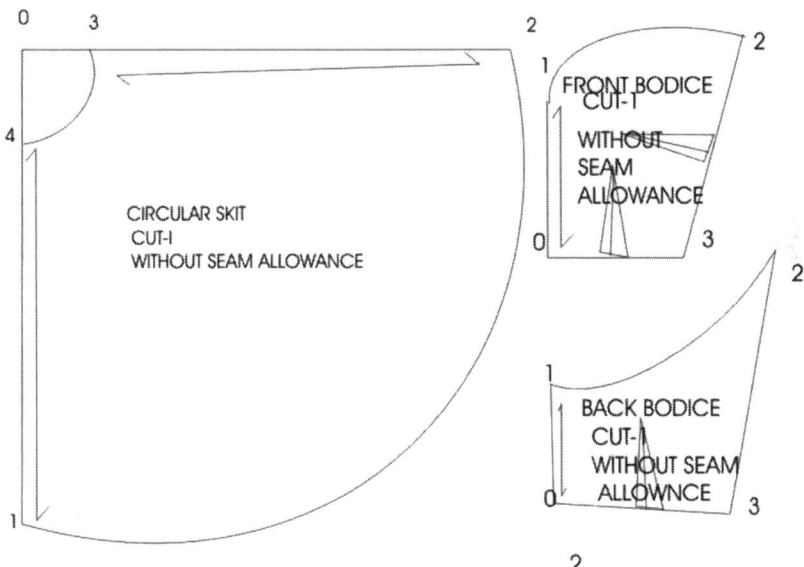

Trim and fabric board

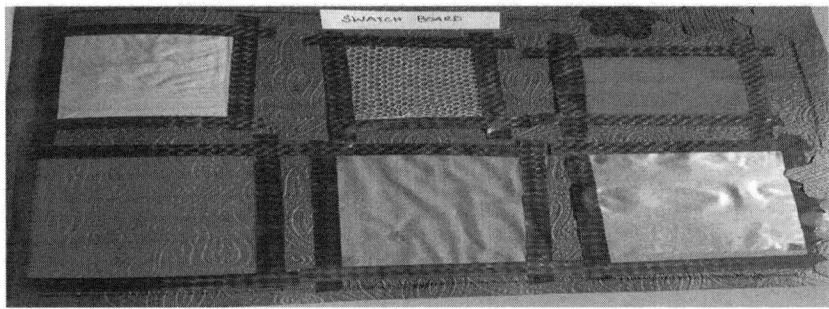

Muslin fit

Sewing a muslin is a great way to test out the construction and fit of a sewing pattern without the risk of wasting good fabric and a lot of time on a garment.

Costing

Materials in meters	Price	Total amount
Satin–7	120	840
Lining material–5	70	350
Surface enrichment material–5	45	225
Lining material for surface enrichment–5	50	250

Trims and accessories

Trims	Price	Total amount
Earrings	200	200
Bangle	100	100
Slipper	1000	1000

Labour cost for stitching: Rs 1000
Total cost of the garment:Rs 3965

Presentation
Collection launch at a fashion runway.

Photos

Conclusion
Cherry blossom theme shows lightness feeling of spring that represents flower symbolism.

3.9   Cheetah

Research—Cheetah

Theme board/Mood board

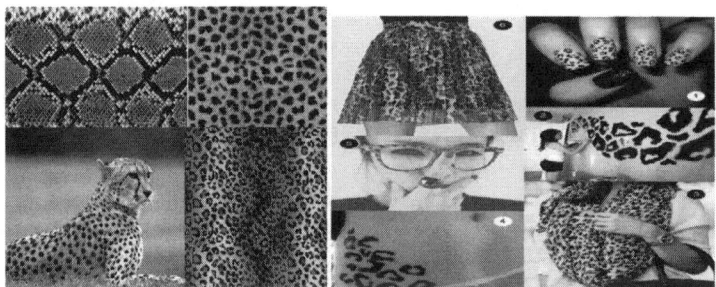

Inspiration board

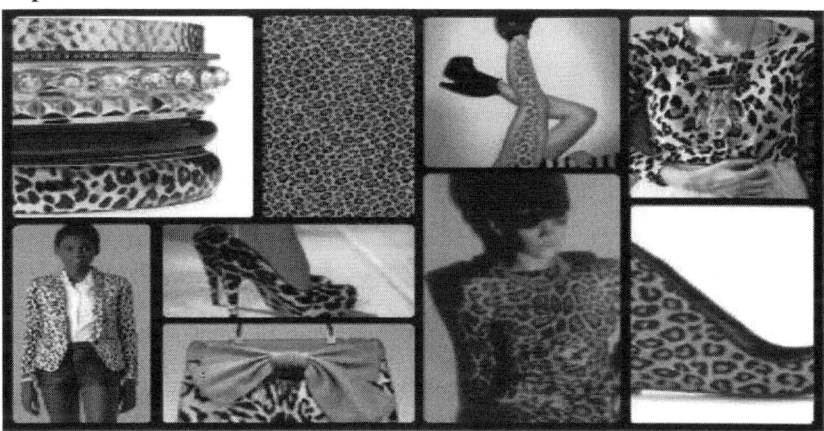

Illustration board

Client board
Age group: 18
Occasion: Formal wear
Profession: Student
Market: National
Description: Inspired by Cheetah with a long tail at the back. With slim fit, Rocky style.

Accessory board

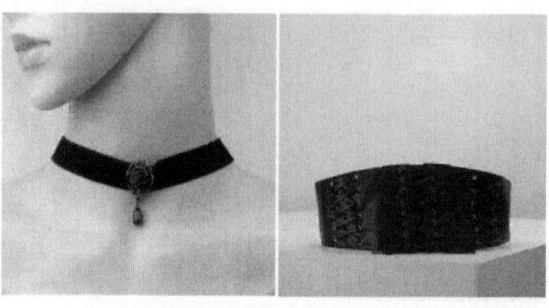

Colour board

Flat sketches board

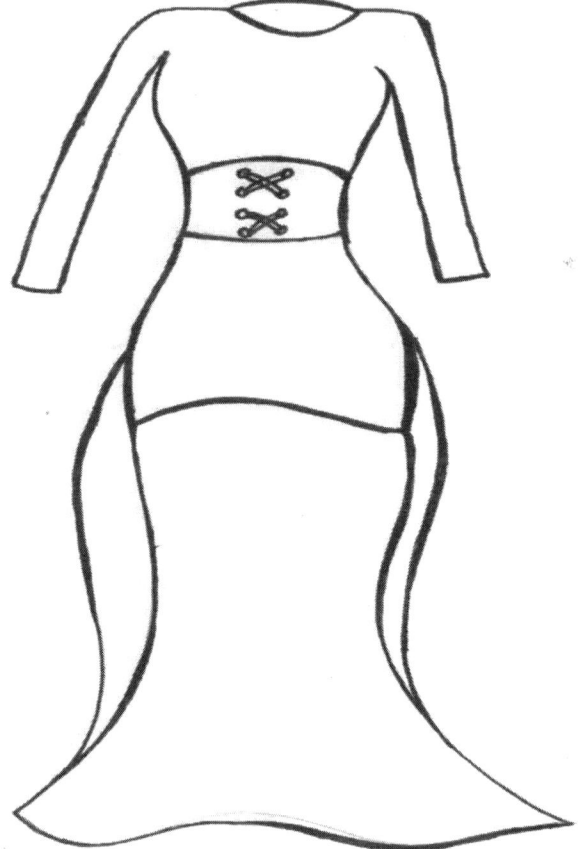

Garment construction pattern details

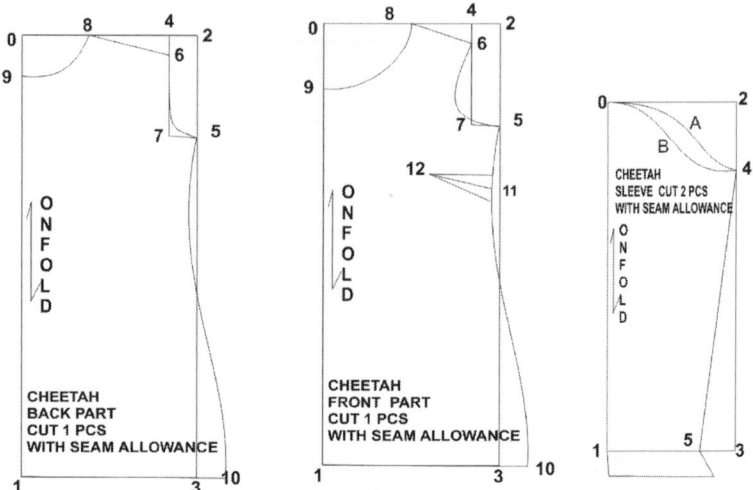

Trim and fabric board

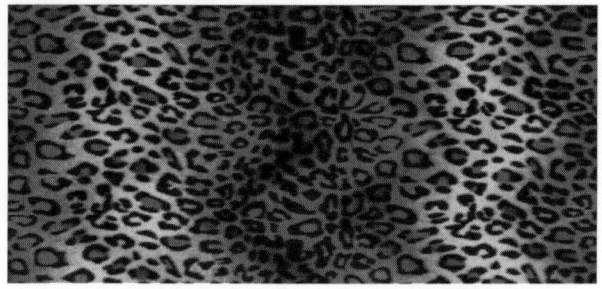

Muslin fit

Sewing a muslin is a great way to test out the construction and fit of a sewing pattern without the risk of wasting good fabric and a lot of time on a garment you ultimately aren't pleased with.

Costing

Materials in meters	Price	Total amount
Velvet material–3	400	1200
Satin material–5	120	600
Leather–1	300	300
Lining–2	70	140

Trims and accessories

Trims	Price	Total amount
Earrings	200	200
Neck accessory	300	300
Slipper	1000	1000

Labour cost for stitching : Rs 500
Total cost of the garment : Rs 4240

Presentation
Collection launch at a fashion runway.

Photos

Conclusion

A new design of cheetah inspired current trends.

3.10   References

1. Kathryn Mckelvey and Janine Munslow, (2008). Fashion Forecasting. Wiley-Blackwell, A John Wiley & Sons, Publication Ltd. ISBN 978-1-4051-4004-1. p137, 150.

2. Armstrong, Helen Joseph, (2006). Pattern Making for Fashion Design, fourth edition. Published by Pearson Education, Inc, Publishing as Prentice Hall, copyright 2006. Original edition, ISBN-978-81-317-2459-0. p 32.

3. Bradford, Julie (2014). Fashion Journalism. Routledge. p129.

4. Jump up Dillon, Susan, (2011). The Fundamentals of Fashion Management. A&C Black. p115.

5. Jump up Caity Weaver, (10 February 2014). "A Minute-by-Minute Account of Fashion Week's Most Harrowing Event. Gawker. Retrieved 21 February 2015.

6. Portfolio presentation for fashion designers-read unlimited e-books and audio books by Linda Tain, Fairchild Publishers, ISBN-10:1563678179, ISBN-B:9781563678172.

7. Martin M. Pegler. Visual Merchandising and Display, Fourth edition. S.V.M Fairchild Publication, New York.

8. Dr. Navneet Kaur. Comdex Fashion Design, volume 11, SDR Printer, New Delhi ISBN-978-93-5004-094-2.

9. Fabrics.net.

10. www.uid.edu.in.

11. www.designersnexus.com.

12. www.fashionising.com.

13. www.ehow.com.

14. www.uen.org.

PART III

4

Need based garment

4.1 Design and construct garment for special people/bed ridden patients

4.1.1 Special people

The clothing for the differently able person (special people) is generally called adaptive clothing or functional clothing. Adaptive clothing is clothing designed for people with physical disabilities, the elderly and the infirm who may experience difficulty dressing themselves due to an inability to manipulate closures, such as buttons and zippers, or due to a lack of a full range of motion required for self-dressing. The clothing specially designed for disabled people must allow more freedom and independence when being worn, but it also has to be fashionable. This is why it is essential that these products be comfortable, appealing to the eye, trendy, easy to put on, take-off, accessible to all those who are disabled, safe and able to adapt to the wearer's physical needs. This particular type of clothing must satisfy the following requirements.

- To provide freedom of movement;
- Be able to keep the handicap under control;
- To provide the required level of safety and comfort;
- To provide the necessary moral and psychological comfort be easy to maintain and cleaned in the washing machine like regular laundry.

4.1.2 Bed ridden patients

Is a form of involuntary bed rest? Medical risks are associated with long-term lying down. The definition of bed ridden generally refers to a person that cannot be out of bed for long period of time if at all due to general weakness, injury, infirmity and illness.

Concerns with dressing or changing a bed ridden patient are as follows:

- Painful process for the patient
- Caregiver struggle with moving a patient

- Caregiver fatigue
- Bed sores
- Lack of circulation
- Incontinent issues.

4.2   Design and construct garment for special people

Loose fitting clothing disability designs allow for freedom of movement and ease of dressing. Women's ¾ length dresses provide modesty and yet afforded easy transfers. The loose fit flows comfortably while seated. Make dressing fit with sleeves, better than velcro, because there is no lint build up from washing and they stay closed and machine washable. Great choice for hospital gowns, seniors who are disabled or wheel chair dependent.

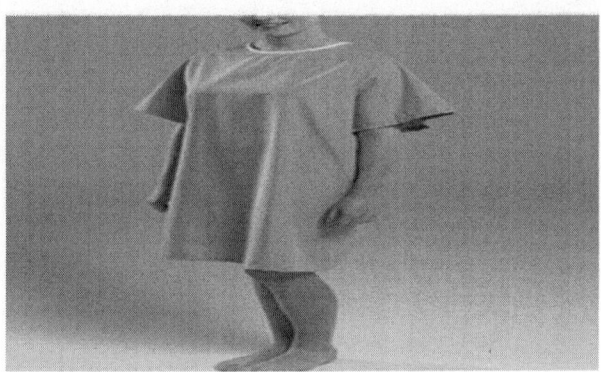

Design of the garment
Aim
To design and construct the special garments for women.

Materials required
Fabric, paper pattern and tool kit.

Measurements details

Full length	:	34″
Waist	:	24.5″
Chest	:	30″
Hip	:	32″
Neck depth	:	3″
Shoulder to waist	:	9.5″

Sleeve length   :   12"

Method of drafting procedure
A–B = Full length (from shoulder to the length you want) plus 1.5
C–D = same as A–B
A–E = 1.5"
A–F = 6"
A–G = 3 plus 0.5"
A–H = 5"
I–J is the stitching line
K–L = 8 or 8.5
M–N Hip round/4 plus 1.5 = J–B, mark-up 1" from J to O from hem curve
O–I = 6–12"

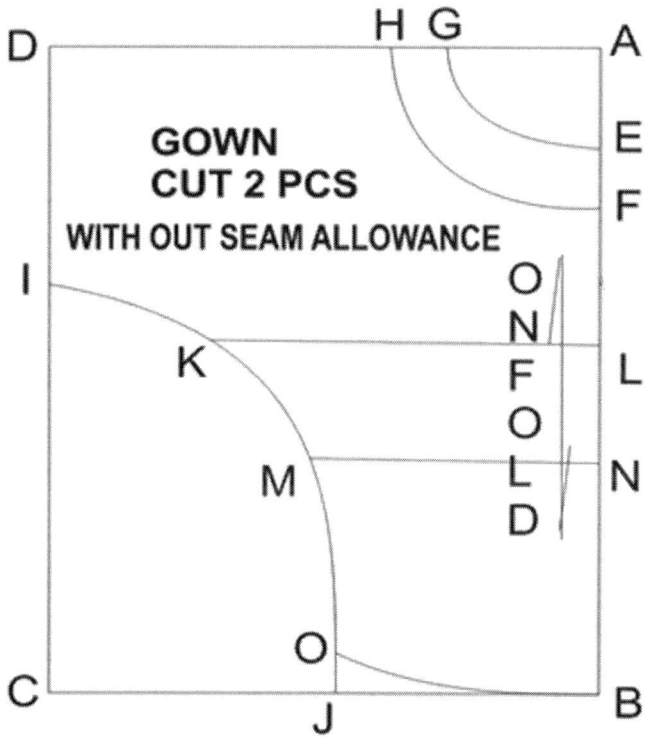

Fabric required for constructing this garment
Approximately 3 m fabrics required for this garment.

Lay out—combination fold

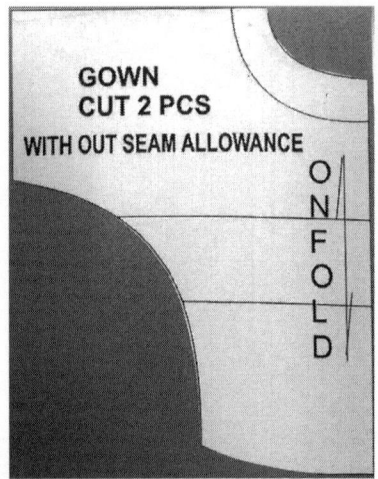

Construction procedure

Step 1: Cut the fabric according to the pattern drafted.

Step 2: The pattern is drafted according to the procedure which means sleeve is come along with the bodice part without any separate cut.

Step 3: Finish neckline with the help of binding and finish the sleeve hem line.

Step 4: Attach the sleeve and side seam.

Step 5: Finally, finish the bottom hem line.

Trims and accessories

Velcro

Cost calculation

i.   Cost of material: Rs 300

ii.   Cost for construction: Rs 150

iii. Total cost: Rs 450

4.3   Design and construct garment for bed ridden patients

Shorts were suspended by the use of velcro located on the side seams. Velcro used for easy touch closure designs for easy self-dressing. The yarns used to make shorts are finely rib-knit for a comfortable fabric that would not fade or lose its shape, even with frequent washing.

Aim
To design and constructed bedridden garment for men.

Materials required
Fabric, paper pattern and tool kit.

Measurements details
Full length   :   16″
Seat   :   33″

Method of drafing
Front
Square lines from 0
1–0 = one-fourth seat plus 7.5 cm (3″)
2–0 = full length plus 1.5 cm (1/2″)
3–1 = one-fourth seat plus 4 cm (1½″)
4–0 = same as 3–1
Join 3–4
5–3 = one-twelth seat
6–3 = one-sixth seat
7–3 = half of 5–3 plus 0.75 cm(1/4″), shape 4–6–7–5
8–2 = 5–1 less 2 cm (3/4″), or
Half of bottom
Shape inside seam 5–8

Back
9–5 = 4 cm (1½″)
10–5 = 2 cm (3/4″)
Join 4–10
11–10 = same as 6–3
Shape fork 11–9 as shown
12–4 = 2 cm (3/4″)
Join 0–12

13–8 = 3 cm (11/4″)
Shape inside seam 9–13
Keep 4 cm (11/2″) above 0–4 and
0–12 for casting
Keep 4 cm (11/2″) below 2–8 and
2–13 for inturns

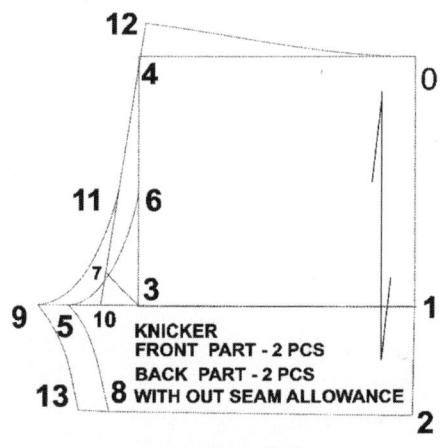

Pattern for knicker

Fabric required for construct this garment
Approximately 1.5 m

Layout —combination fold

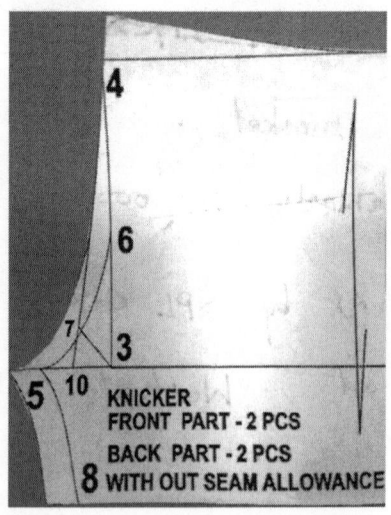

Construction details
Step 1: Cut the fabric.
Step 2: Join the four pieces.
Step 3: Join the waistband with elastic.
Step 4: Attach the Velcro.
Step 5: Trim the extra thread and fabric.

Trims and accessories
Velcro

Cost calculation

i.   Cost of material: Rs 300
ii. Cost for construction: Rs 80.
iii. Total cost: Rs 380

4.4   References

1. Apparel construction, 3G C learning, copyright-2014 by 3G ELearning, FZLLC, UAE, ISBN-978-93-5115-123-4. p 97, 99IO.
2. Deepti Pargai, Manisha Gahlot, Anita Rani, (2015). Designing and Construction of Functional Clothing for a Child with Disability: A Step towards the Social Responsibility. The International Journal of Social Sciences and Humanities Invention. 2(9):1534–1541. ISSN: 2349-2031.
3. Curteza Antonela, Cretu Viorica, Macovei Laura, Poboroniuc Marian, (2014). Designing Functional Clothes for Persons with Locomotor Disabilities. AUTEX Research Journal, 14(4): 281–289.
4. Kobayashi, Erin (2007). If the hospital gown fits Toronto Star. Retrieved 30 August 2014.
5. Frey, Rita, Cooper, Lisa Shearer, (1996). Introduction to Nursing Assisting: Building Language Skills. Delmar Learning. p 264. ISBN 0-8273-6233-1.
6. Gown: noun. Longman Dictionary of Contemporary English. Pearson ELT. Retrieved September 20, 2013. 3: a long loose piece of clothing worn in a hospital by someone doing or having an operation.
7. Carter, Pamela, J. (2007). Lippincott's Textbook for Nursing Assistants: A Humanistic Approach to Caregiving. Lippincott Williams & Wilkins. p 378. ISBN 978-0-78176-685-2.
8. Rosdahl, Caroline Bunker, Kowalski, Mary T, (2008). Textbook of Basic Nursing. Lippincott Williams & Wilkins. p 499. ISBN 978-0-78176-521-3.
9. Simple techniques slash hospital infections: meeting. Reuters. March 21, 2009.
10. Blackwell, Tom, (2014). Canadian study dresses down hospital gowns. National Post.
11. https://www.silverts.com/clothing-for-bedridden-patients/.
12. Adaptive Clothing – Information and Availability". Disabled World. 20 March 2014.

PART IV

Garment accessory making

(Design and construct gloves/hat/socks/veils/belt/bow/tie/bags)

5.1   Introduction

Accessories are a means to express different moods and create different looks with similar kind of ensemble. They lead the viewer's eyes to the assets of the wearer and work towards co-ordinating the outfit, with the help of dangling earrings, chunky necklaces and hanging belts, etc. The accessories can make outfits suitable to be worn for different occasions. The selection of accessories provides personalizes effect to the garments. Accessories are added to the look of the outfit, as expensive accessories make the whole ensemble look richer and sophisticated. The design of the accessories must be suitable with new fashions and also with the cloths already owned. So the fashion accessories manufactures regularly should forecast the changes in the ready-to-wear market and should produce styles that are innovate, leading, compatible and impressive. The fashion accessories industries should be extremely reactive to fashion change and fast to understand the future trends.

An accessories often so small, and yet so powerful. Pairing the right accessories with an outfit will take your look up a notch. Your simplest sheath can morph from basic to sexy as fast as you can reach for a belt, bow, bags, etc. Mixing and matching these ensemble extras is a fast and easy way to show-off your personal style.

Fashion accessories are decorative items that supplement and complement clothes, such as gloves, handbags, hats, belts, bows, ties and socks. Accessories can add colour, style and class to an outfit, and create a certain look, but they can also have a practical function, handbag can be used to carry small items such as cash, hats, protect the face in bad weather, and gloves keep hands warm.

5.2   Types of garment accessories

Various kinds of accessories are used on garments, some are part of the garments such as buttons, zippers, interlining, etc., while others are used for decorating and enhancing the product appearance such as sequins, embroidery, etc.

Normally garment accessories can be classified in three ways as follows:
1. Garment accessories/basic accessories
2. Decorative accessories
3. Finishing accessories

Finishing accessories
Some of the finishing accessories are as follows:
- Gloves
- Hat
- Socks
- Veils
- Belt
- Bow
- Tie
- Bags

5.3   Gloves

5.3.1   Introduction

A glove is a garment covering the whole hand. Gloves may be used as separated sheets (or) openings for the thumb and each finger. Gloves protect and gives comfort to our hands against cold (or) heat. In gloves there is an opening but no covering sheath for each finger, they are called fingerless gloves. Woven, knitting, felted, wool, leather, rubber and latex fabrics are used for preparing gloves. Today gloves are manufactured by sewing machines (or) gloves can be made by hand either as the decorative stitch. The decorative stitches used in gloves are flat stitch, running stitch, round stitch, and knotted stitch.

Gloves have been used to keep their hands warm, to insulate from heat, to protect from electric shock, to grip, to fight and to decorate. Though many materials can be used in making a glove but the most popular material used is leather.

5.3.2   Material used for making gloves

Cotton, cotton blends, polyethylene, polyvinyl chloride, natural and synthetic rubber, Kevlar and Teflon.

5.3.3   Types of gloves

Leather gloves, food service gloves, driving gloves, gardening gloves, disposable gloves, fabric coated gloves, chemical resistant gloves, surgical gloves and military gloves.

5.3.4   Glove size

With the help of measuring tape measure horizontally around palm of hand when open (excluding thumb). Then add your ease allowance. Thumb also measured separately with ease allowance. The measurement will be in inches.

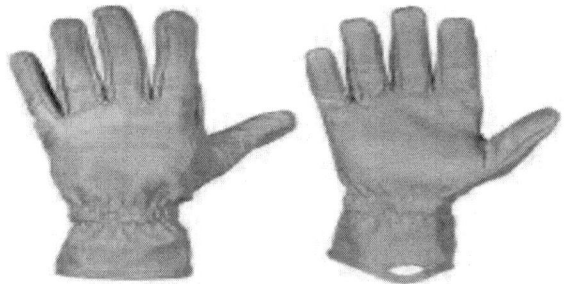

5.3.5   Simple method of making gloves without measurements

To make a pair of gloves the following steps are carried out.

Step 1
Lay your hand with your thumb hanging-off on a folder piece of pattern drafting paper. Trace around your hand, allow extra seam allowance. Mark all drafting details.

Step 2
Cut two pieces, one for your left hand and other one for your right hand.

Step 3
Measure and cut three fabrics (one for interfacing) for each hand to sew in between the fingers.

Step 4
Sewing them together, to get more freedom of movement cut the thumb separately and sewing it on at an angle matching.

5.4   Hats

5.4.1   Introduction

A **hat** is a head covering which is worn for various reasons, including protection against weather conditions, ceremonial reasons such as university graduation, religious reasons, safety, or as a fashion accessory. In the past, hats were an indicator of social status. In the military, hats may denote nationality, branch of service, rank or regiment. Police typically wear distinctive hats such as peaked caps or brimmed hats, such as those worn by the Royal Canadian Mounted Police. In the world of hats, there are plenty of styles and types from which to choose. Each hat provides a different type of function and style depending on your wants and needs. Hats can be as simple as something that looks good on you, like a fashionable or something that serves a primary purpose, like a drape hat for a hiking trip.

5.4.2   Materials used for hat

Camel hair: Specialty fibre woven into hair cloth, generally coarse and inflexible. When blended with wool or taken from the pure undercoat, camel hair becomes quite plush and soft. Haircloth is taken from the undercoat when camels molt in the warmer seasons.

Cashmere: Sometimes referred to as Pashmina. Taken from the Cashmere goat, the wool is fine in texture, strong, soft, and light. It is extremely warm when used in garments.

Murino wool: Hats in the Belfry's own term for the fur felt-like finish of a hat. Through the steaming and brushing process during production, murino-finished hats have the soft feel and look of fur felt, without harming any animals in the process.

Silk: Natural protein fibres woven into textiles for production. The naturally strong absorbent qualities make silk comfortable to wear in warm weather. The low conductivity keeps warm air from escaping your head, making it also desirable in the winter.

Worsted wool: Long, fine staple wool that is spun to create worsted yarn. The wool is brushed before woven, creating a finish that is extremely soft and light weight, but just as warm as regularly prepared wool. Worsted wool is more durable than regular wool felt.

5.4.3   Types of hat

1. Bobble hat
2. Cow boy hat
3. The bucket hat
4. Porl pie hat
5. Sailor hat
6. Top hat
7. Fedora hat
8. Fez hat

9. Gambler hat
10. Safari hat, etc.

Design of the hat

Wide-brimmed summer hat

5.4.4   Material required

- Kraft paper
- Glue or scotch tape
- Tape measure
- Compass
- Approx. 1 yd of Pellon® fleece interfacing, 45″ wide
- Approx. 1 1/2 yds of bottom-weight or upholstery fashion fabric, 54″–60″ wide
- Approx. 1/2 yd of lining fabric, 45″ wide
- Approx. 3/4 yd of grosgrain ribbon, 2″ wide
- Co-ordinating thread.

5.4.5   Measurements

Begin by measuring the circumference of your head. Wrap the tape measure around the back of the head, over the ears, and high-up on the forehead. For most people, this will be approximately 22½″. Add 1″ to your head measurement. This is the head fitting measurement. (This number may vary due to hairstyles as well. Most people will pull back the hair or put it in an up-do to wear the hat properly.)

Next, you need to determine the radius (**R**) you will use to create a circle.

Use the formula, $C = 2\,n\,R$, where $n = 3.14$.

E.g. $23.5 = 2 \times 3.14 \times R$

$R = 23.5$ divided by 6.28

$R = 3.75$.

For a head fitting measurement of 23½″, you need a radius of 3¾″.

5.4.6   Pattern

Crown

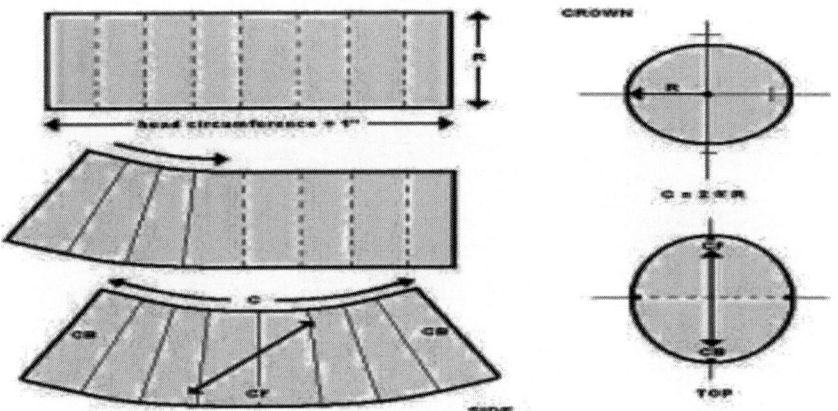

Draw a rectangle equal to the head fitting measurement X the radius amount.
Divide the rectangle into eight equal parts.

Slash each line from the top of the rectangle to almost the bottom edge.

Overlap each section left of centre by ½″ at the top to nil at the bottom and glue down.

Repeat for each section right of centre.

Trace this new shape onto a new sheet of paper and blend the curved top line smooth.

Add ½″ seam allowance around the perimeter. Label the two short ends, centre back (**CB**) and the centre line, centre front (**CF**).

Draw a **grain line** 45° to the centre line. This is the side section of the crown.

Measure the new curved seam that you have created. Record this measurement.

Using the formula again, determine **R** for the top of the crown.

If you measure the circumference of the top section, it should equal the curved seam measurement of the side section.

(**Adjustment Tip**: *Increasing or decreasing the circle by 1/8" all around will change the circumference by approximately ½"*).

Use a compass and draw a circle using the amount for the radius (**R**).

Divide the circle into quarters. At the top and bottom of the circle extend out ½". At the sides of the circle, reduce inwards by ½". Join these points with a smooth arc blending into each point to create an oval. Re-check the circumference of the oval. It should still equal to that of the new curved seam on the side section.

Add ½" seam allowance to the circumference and draw a straight grain line from **CF** to **CB**.

This is the crown top.

Brim

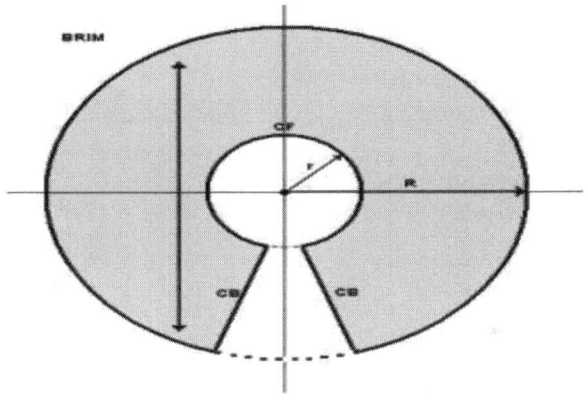

Fold a large piece of paper in half, then half again; open out flat.

Draw the two intersecting lines. Place a point in the centre of the intersection.

Using a compass, draw a circle with a radius (*r*) based on the head fitting measurement plus 1".

From the same point, draw a second circle with a radius (**R**) equal to three times *r*.

This is the brim width. (*You may increase the width of the outer edge if you desire*).

The inner circle must equal the head fitting measurement. Starting at the top of the circle, measure half the head measurement to the left of centre and repeat on the right of centre. Mark a point at each interval. Square out from this point to the outer edges. Label these edges, CB. This is the interfacing pattern.

Trace this brim shape onto a new sheet of paper. Add ½" seam allowance all around the brim pattern. Draw a straight grain line parallel to the centre line. This is the upper and lower brim patterns.

Cutting

Cut the crown side **1X**; lining **1X (on bias grain).**
Cut the crown top **1X**; lining **1X (on straight grain).**
Cut the crown side interface **1X (on bias grain).**
Cut the brim **2X (on straight grain).**
Cut the brim interfacing **1X (on straight grain).**
Cut 1 strip of self-bias 2″ wide; length to equal outer edge of brim + 2″ **(on bias grain)**

Assembly

On each brim piece, pin and baste **CB** seam with right side together. Stitch seam using a ½″ seam allowance and press open.

On the brim interface piece (Pellon®), butt the two **CB** edges together and zigzag stitch it together.

Sandwich the interfacing between the two brim pieces with face sides up, lining up the **CB** seams. Baste together.

Do rows of topstitching parallel to the outer edge, ½″ apart, starting from the inner edge. (*I use the width of the presser foot as a spacer.*) The more stitching you put, the stiffer the brim.

Bind the outer edge of the brim with the bias cut strip of self-fabric, beginning from the **CB** seam, folding in ½″ to bind, and then turning under the end to encase the starting point. Press edge flat.

Baste the side interface to the wrong side of the side section.

With right sides together, pin and baste the **CB** seam. Stitch seam using a ½″ seam allowance and press open. Grade seam allowances.

Divide top edge of crown side and the crown top into quarters. Match up these points, pin and basting with right sides together. Machine stitch the side section to the top of the crown, using a ½″ seam allowance. Trim and grade the upper seam allowance and interfacing only.

Repeat the same procedure for the hat lining.

With wrong sides together, place lining into crown, aligning with **CB** seam and baste seam allowances together. Turn right side out.

Now, insert crown into centre hole of brim. Match up **CB** seams and pin and baste crown to brim. Machine stitch using a ½″ seam allowance. (*Clip seam allowance if there is any buckling.*)

Sew in grosgrain ribbon as a "sweatband". This should equal the head fitting measurement less the inch. Lay the ribbon on top of the seam allowance and edge stitch along the ribbon edge. Overlap at **CB**. Trim the seam allowance with pinking shears. Turn the ribbon up into the crown to conceal the raw edge of the seam allowance and tack to **CB** seam with hand stitching.

5.5   Socks

5.5.1   Introduction

A sock is an item of clothing worn on the feet and often covering the ankle or some part of the calf. Some type of shoe or boot is typically worn over socks. In ancient times, socks were made from leather or matted animal hair. They help protect our feet, keep them warm and prevent out feet from getting blisters while wearing shoes. In many ways socks are the unsung heroes of your wardrobe.

5.5.2   Materials used in socks

Cotton, wool, nylon, polyester, acrylic, spandex, polypropylene, olefin, profilen, Gore-tex, Kevlar, X-static.

5.5.3   Types of socks

1. Ankle length socks
2. Quarter length socks
3. Crew length socks
4. Mid-calf length socks
5. Calf length socks
6. Knee length socks
7. Thigh high socks
8. Slip on paddings

5.5.4   Step by step procedure for making socks

Step 1

Measure this length on the cardboard and draw a line from the beginning to the end of the measurement. Measure from your toe to your ankle and mark the spot where your ankle hits on the cardboard.

Step 2

Measure around your ankle and divide the measurement in half. Measure the halved amount from your ankle mark the length on the cardboard.

Step 3

Measure from your toe to your heel and mark this spot on the cardboard as well. If your socks will go above your calves, mark any spots where your leg width fluctuates and measure down from the length line. Fluctuates and measure down from the length line.

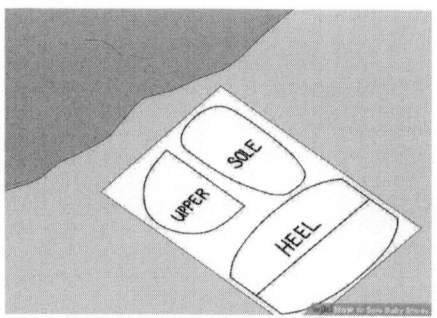

Step 4

Using the marks of your heel and ankle, draw another line to create a pattern for your sock. Remember to look at the shape of the foot while completing this step to be accurate. Cut out the pattern and fold your chosen fabric in half.

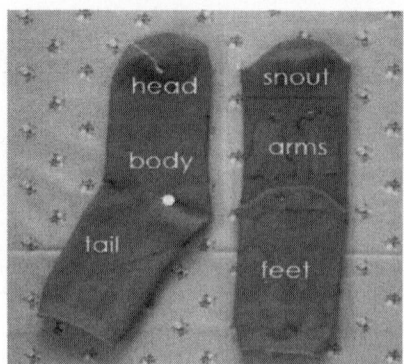

Step 5

Thread the needle and tie the ends together. Place your cardboard pattern against the folded edge of the fabric and trace the pattern.

Step 6

Cut the fabric. Sew the length of the pattern but do not sew the toes shut.

Turn the folded fabric so they seam runs up the back of the sock. Sew straight across the toe line.

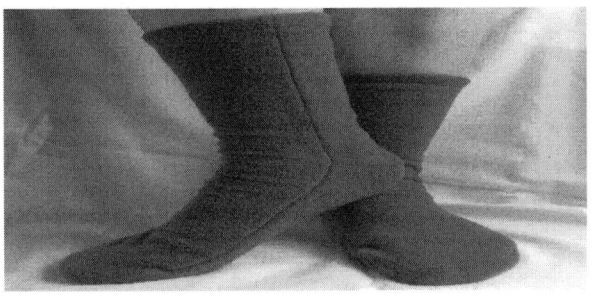

5.6   Veils

5.6.1   Introduction

A **veil** is an article of clothing or hanging cloth that is intended to cover some part of the head or face, or an object of some significance. The veil is usually the last finishing touch and can be one of the most defining aspects of your bridal appearance. Once you decided to wear a veil, you will need to consider how it works with your dress, face, shape, hairstyle and wedding location. Whether you want a modern or classic wedding, every bride deserves a gorgeous wedding veil to walk down the aisle in and knowing that you made it yourself makes it more special.

5.6.2   Materials required for veil

Tulle fabric, hair comb, sewing threads, sewing needles, scissors and measuring tape.

5.6.3   Types of veils

1. Birdcage veil
2. Blusher veil
3. Mantilla veil
4. Tiered veil

5. Fingertip veil
6. Ballerina veil
7. Cathedral veil and
8. Elbow veil.

5.6.4   Measurements (veil lengths) required for veil

1. Shoulder–22″
2. Elbow–25″
3. Waist–30″
4. Mid-hip–33″
5. Hip–36″
6. Fingertip–45″
7. Waltz–54″
8. Ankle–70″
9. Chapel–90″
10. Cathedral–108″

5.6.5   Step by step construction procedure for veil

Step 1
Cut one piece of tulle to your desired length × 3 yards wide.

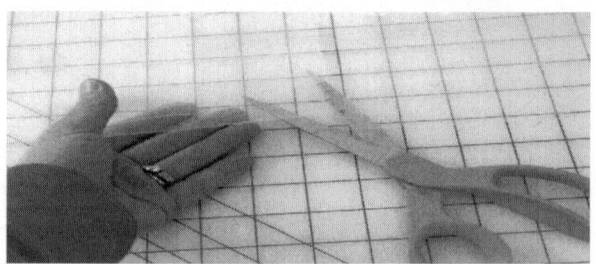

Step 2

Sew a line of gathering stitches along the top edge of the veil. A gathering stitch is a long basting stitch, with the ends not stay-stitched.

Step 3

Pull bobbin threads–Gather the top of the veil by pulling the bobbin threads on either side. Gather the top of the veil as much you can, until it is about the same size as your hair comb.

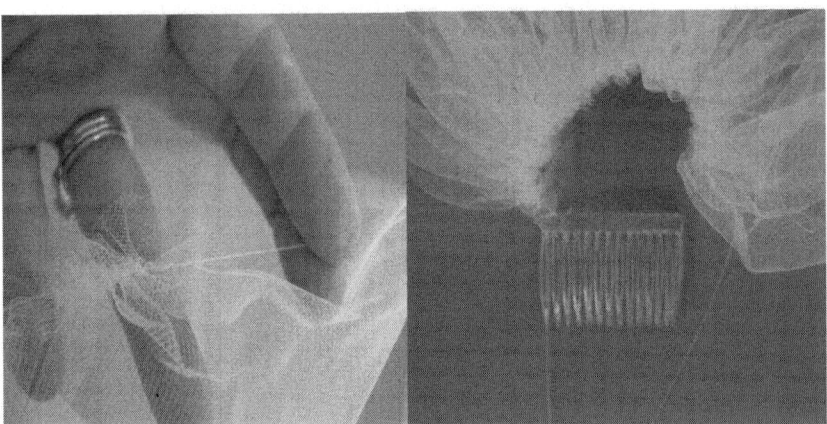

Step 4

Sew gathered top–Use a regular stitch to sew over the top gathered edge, pushing the gathered veil as close together as possible while going through the machine. Make nice tight gathered top, this will secure your gathered veil together.

Step 5

Create clean line. Cut-off the excess gathered ends to make the tulle nice clean line.

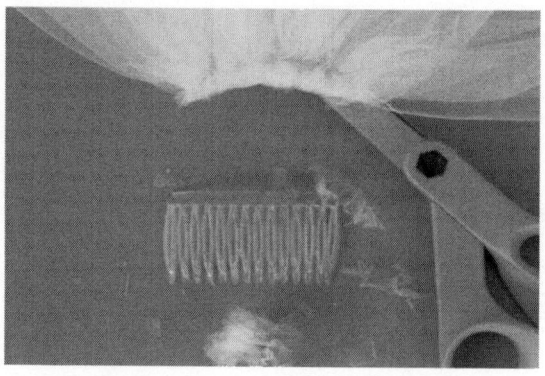

Step 6

Sew to comb–Flip over the comb so that it is wrong side against the veil. Line up the top of the veil with the top of the comb. Use a thread and needle to secure the gathered end of the veil around the top of the comb.

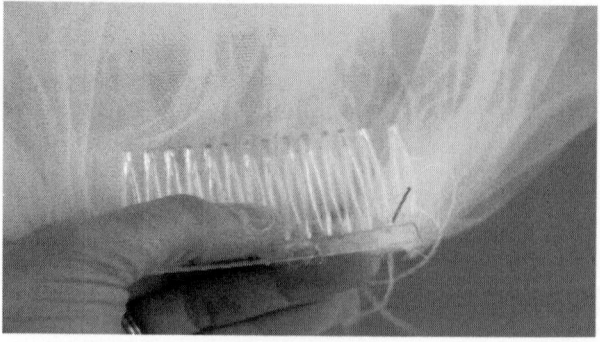

Step 7

Sew entire comb–Make sure you loop through the entire top end of the comb with the thread.

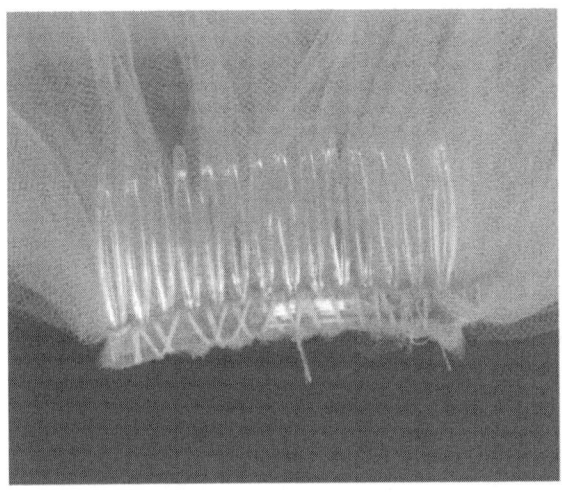

Step 8

Flip over–Flip the comb back over. Your veil should now cover the stitches.

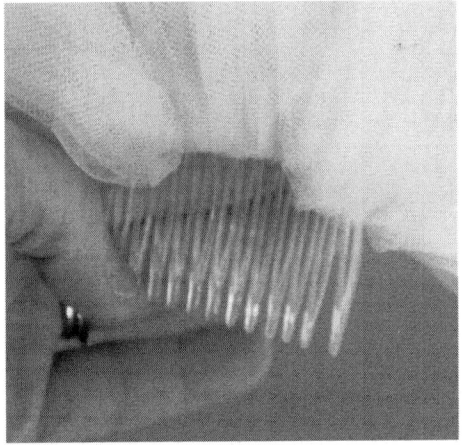

Step 9

Add ribbon–Add a nice silk ribbon to the bottom of the veil. Sew the ribbon right side up, directly on to the right side of the veils hem. Sew the ribbon directly on top of the tulle, along the entire bottom edge. You can either use fray check on the ends of the ribbon under ¼" when sewing. Now push the comb in to your hair to secure and wear.

5.7   Belts

5.7.1   Introduction

Belt is worn around the body usually on the waist area to secure lower garments, such as trousers and skirts and with full garments like dresses. Although most people wear belts for their practical purpose, it helps to improve the overall personality of the wearer. Decorative or functional item worn circling above below or at the natural waistline. Also worn over the shoulder in military fashion. In the past and in poetic writings also called a girdle. A belt is a flexible band (or) strap, typically made of leathers (or) heavy cloth and worn around the waist. A belt supports trousers (or) other articles of clothing and holds the garment in place without slipping down.

5.7.2   Material used for making belts

Fabric, leather, canvas, rubber, etc.

5.7.3   Types of belts

Casual belts, dress belts, designer fashion belts, uniform belts and work wear belts.

5.7.4   Material required for making belts

- A non-stretchy material

- Belt buckle from an old belt or craft store
- Scissors
- Sewing needle and thread or machine
- Iron and ironing board.

5.7.5   Measurement required for belt

For this belt, measure around the part of your waist where the belt will be worn. Some people like to wear belts on their waist, while others like to wear theirs at the hip. Be sure to measure where you will be wearing the belt. Add 5″ to the waist measurement. How thick do you want the belt? Try 1″ for a skinny belt, or 4″ for a fat belt. Keep in mind that thicker belts won't fit through pant loops. Add 1″ to the width of your belt for seam allowance.

5.7.6   Step by step procedure for making belt

Step 1–Measure and cut fabric

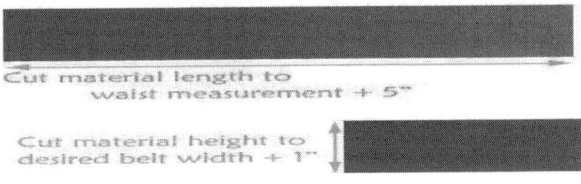

Step 2–Fold and stitch the ends

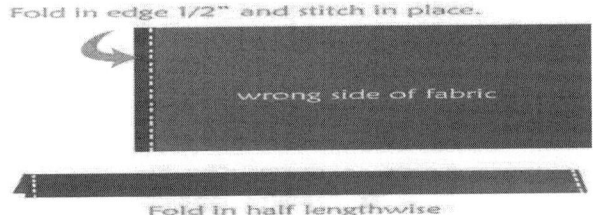

Fold in the edges and stitch. Fold in the ends of the belt a 1/2″ and press with your iron. Before you press, test a small portion of the fabric to be sure that the fabric you are using won't burn under the iron.

For fabrics that may be sensitive to heat (synthetic fibres like vinyl's and plastics) try laying a clean towel between the iron and the fabric while you press.

Step 3–Fold in the bottom, press and sew

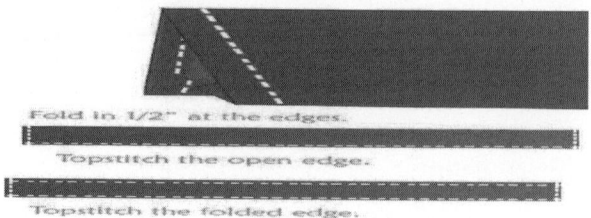

Topstitch both the bottom and top of the belt. Fold up the bottom of the belt a 1/2″ on each edge and press. Top stitch the bottom together a 1/8″ from the edge. For a finished look, press and top stitch along the top of the belt as well.

Step 4–Attach the buckle

Fold fabric over the middle of the belt and stitch. A large variety of belt buckles can be readily found at your local craft store, or you could use a buckle from an old belt that has seen better days.

For this particular belt, we are using a buckle that does not require holes in the belt. If you want holes in your belt, simply use a grommet punch to add holes. Fold over the end of the belt material around the middle rung of the belt and stitch in place. Back-stitch and tie off the edges of string to ensure that this stitch does not come undone.

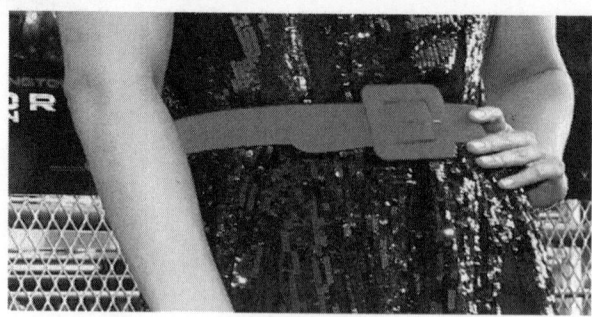

Bold belts can change an outfit. Use a brightly coloured belt with a black or white outfit of an interesting or smooth texture. Pair a patterned belt with muted solids. If you are adding grommet holes, measure so that they are evenly spaced.

5.8   Bows

5.8.1   Introduction

The **bow tie** is a type of traditional necktie. A modern bow tie is tied using a common shoelace knot, which is also called the **bow knot**. It consists of a ribbon of fabric tied around the collar of a shirt in a symmetrical manner so that the two opposite ends form loops. The most traditional bow ties are usually of a fixed length and are made for a specific size neck. Sizes can vary between approximately 14″ and 19″ as with a comparable shirt collar. Fixed-length bow ties are preferred when worn with the most formal wing-collar shirts, so as not to expose the buckle or clasp of an adjustable bow tie. Adjustable bow ties are the standard when the tie is to be worn with a less formal. The bow tie knot is used to tie and is worn to give you a formal and elegant appearance.

5.8.2   Material used for making bows

Cotton, cotton blends, silk, polyester, and velvet.

5.8.3   Types of bows

1. The self-tie
2. The pre-tied
3. The clip-on

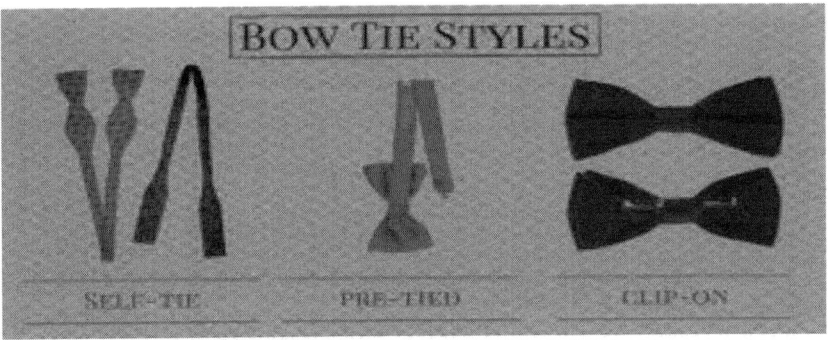

5.8.4   Measurements used for making bow

1. Fabric 10″ × 12″
2. 6″ length of ribbon–3/8″ wide to ¾″ wide for this size bow.

5.8.5   Step by step procedure for making bow

Step 1

Cut a rectangle of fabric twice the width and twice the length of the size of your desired bow. My 10″ × 12″ rectangle of fabric will create a bow that is roughly 5″ × 6″.

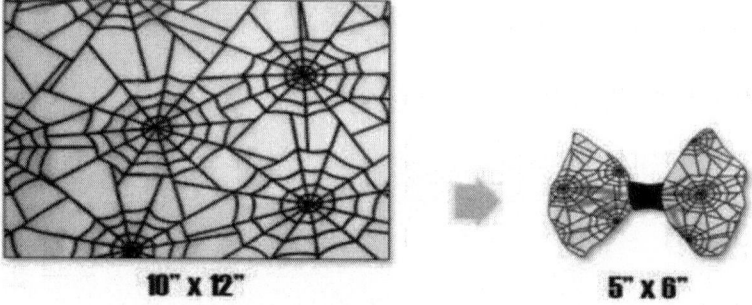

10″ x 12″   **5″ x 6″**

Step 2

Fold the rectangle in half, right sides together, lining up the longest sides. Pin if necessary.

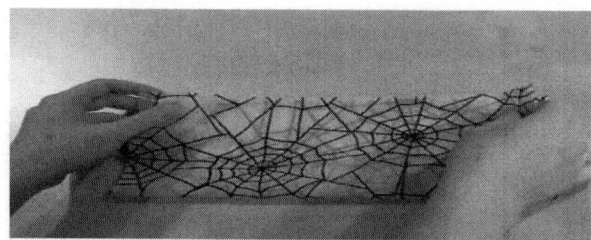

Step 3

Sew along the long edge, using the seam allowance. Use a regular straight stitch on your sewing machine.

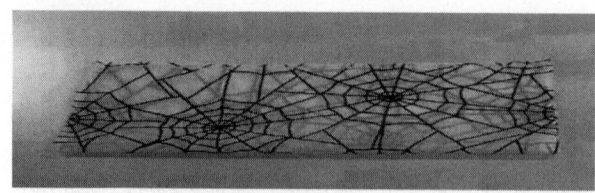

Step 4

Now you have a tube of fabric. Turn the tube right side out, and lay it flat with seam facing up. Adjust as necessary to centre the seam as best as you can.

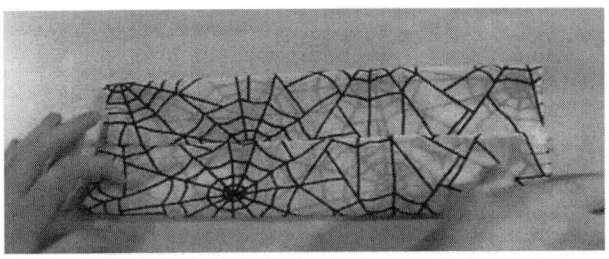

Step 5

Flip the fabric over, so that the seam is on the bottom.

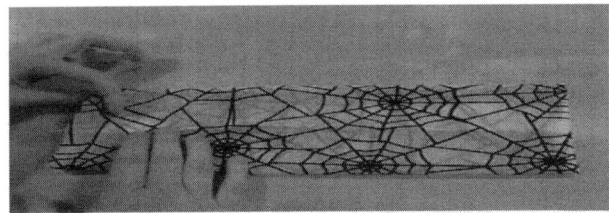

Step 6

Fold the rectangle in half, matching up the unfinished edges. Pin, if necessary.

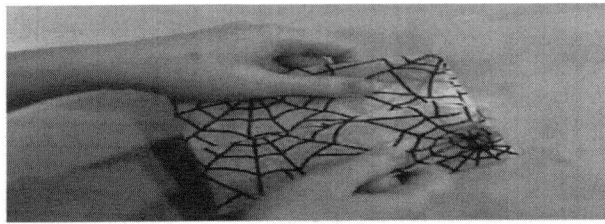

Step 7

Sew the seam

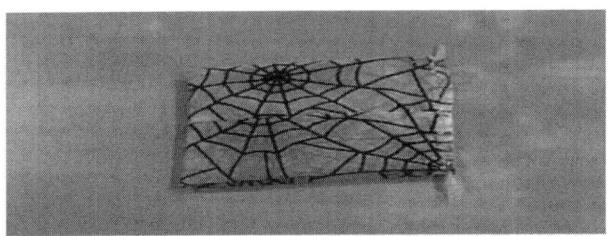

Step 8

The result is now a smaller tube. Turn the tube right side out and centre the seam.

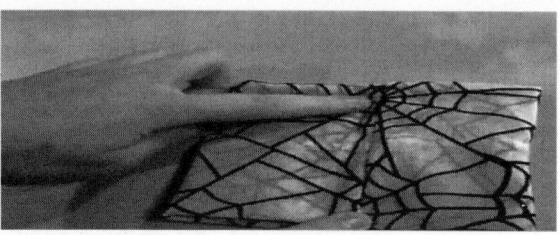

Step 9

Flip it over so the seam is facing down. The tube "opening" should face north and south, at this point. Starting at the centre, fold the rectangle accordion style.

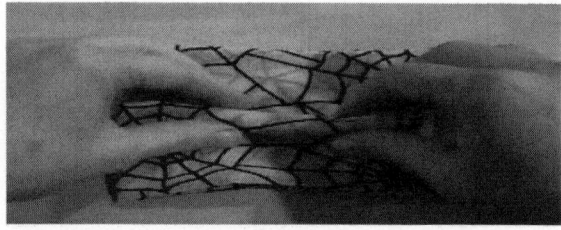

Step 9.1

The photo shows the rectangle after folding.

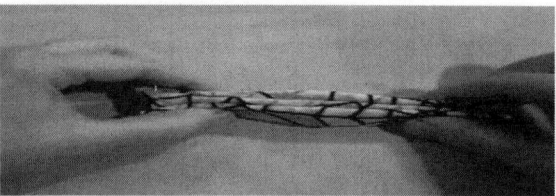

Step 10

Use a binder clip to hold the folds in place at the centre of the bow. You could also use a clothespin or you could baste the centre in place. Hand sewing would probably be easiest, considering all the bulky layers.

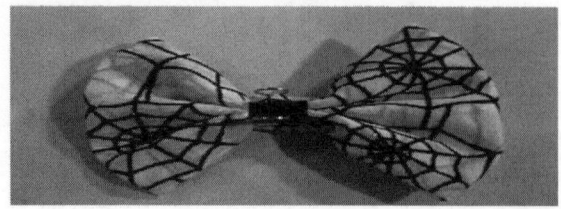

Step 11

Squirt a glob of hot glue on the end of the ribbon.

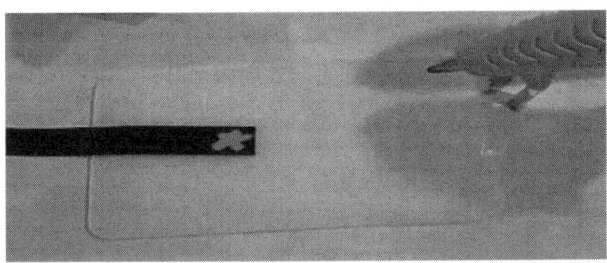

Step 12

Place the back of the centre of the bow on the glue. Let the glue set.

Step 13

Squirt about a 1″ line of glue along the ribbon.

Step 14

Remove the clip, using the other hand to keep the bow folds in place.

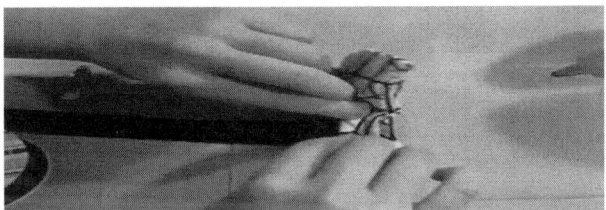

Step 14.1

Roll the bow in to the glue, and keep rolling until the ribbon wraps all the way around the bow.

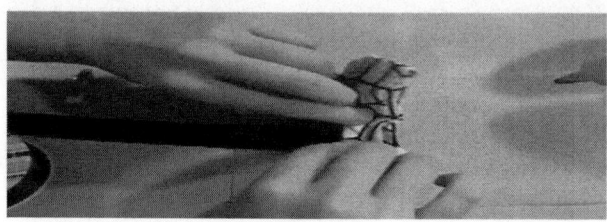

Step 15

Hold the ribbon in place until the glue sets.

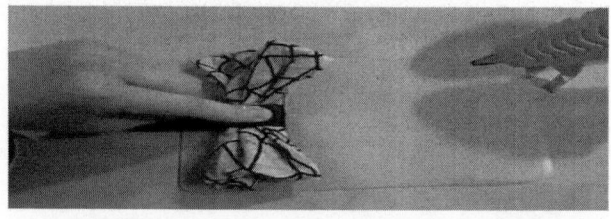

Step 16

Trim the ribbon, leaving a 1.5″ tail.

Step 17

Slide the ribbon through the fastener.

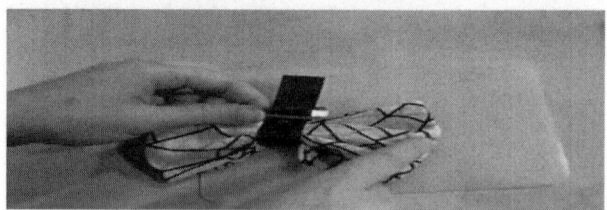

Step 17.1

Squirt a small dot of glue on the back of the bow.

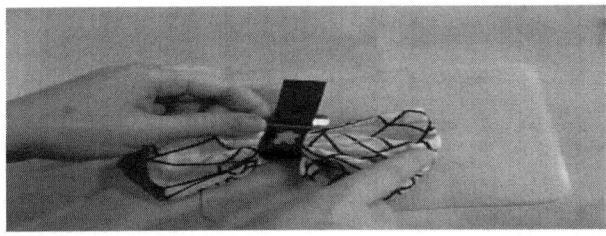

Step 18

Press the fastener and ribbon in to the glue. If possible, fold the raw edge of the ribbon underneath first, so the raw edge is then sealed by the glue. Allow the glue to set.

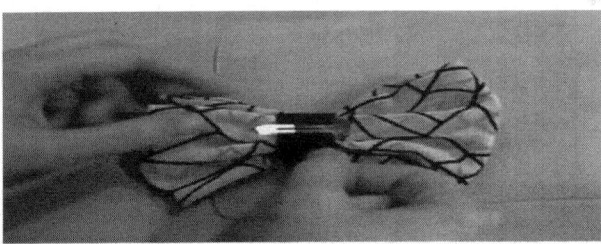

Step 19

The final bow.

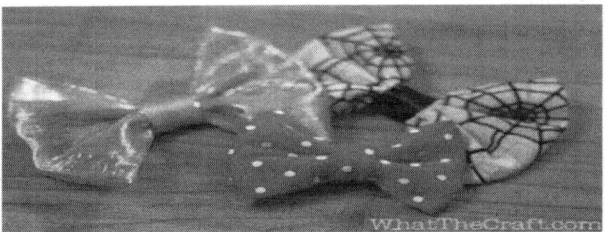

5.9   Ties

5.9.1   Introduction

A **necktie**, or simply a **tie**, is a long piece of cloth, worn usually by men, for decorative purposes around the neck, resting under the shirt collar and knotted at the throat. Ties are gaining increased popularity as trendy accessories that can be worn outside the traditional office setting. Ties can be made from just about any type of fabric and are easy for anyone to create.

5.9.2   Material used for making tie

Cotton, silk, rayon, polyester, linen and denim.

5.9.3   Types of ties

Bowtie, four in hand necktie, the seven fold tie, shiny necktie, western bowtie, bolo tie, cravet tie and neckerchief.

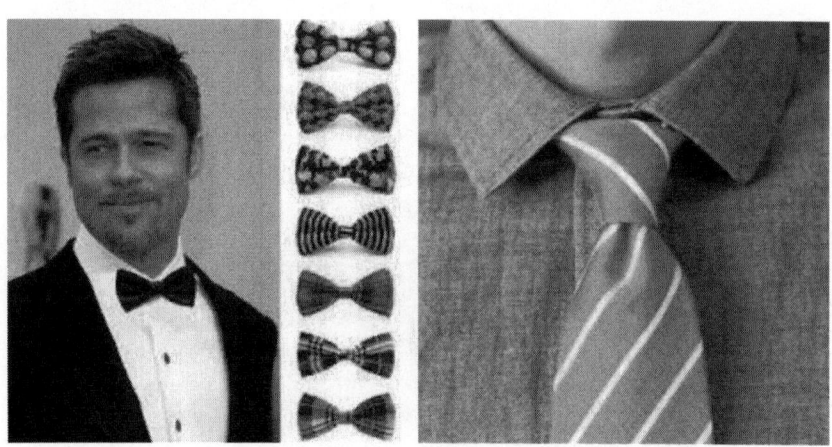

Bowtie   **Four in hand necktie**

The seven fold tie   **Shinny necktie**

Western bowtie

Bolo tie

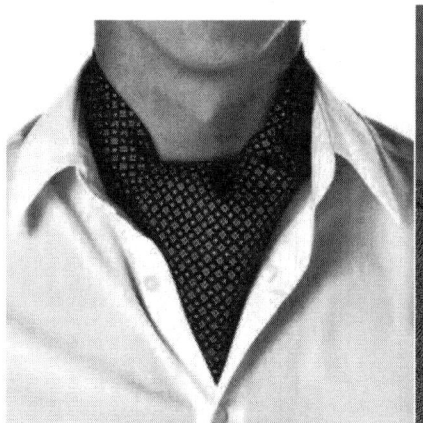

Cravet tie

Neckerchief

5.9.4   Material required for constructing tie

- Cardboard or craft paper for drafting the pattern
- Straight edge ruler
- Marking pen/pencil
- Rotary cutter or very sharp long shears
- Silk pins
- Needle for hand sewing
- ¾ yards of fabric for the tie base, 45″ or 60″ wide

- ¼ yard of lining material
- 1¼ yard of medium weight sew-in interfacing
- Matching thread.

5.9.5　Measurements required for constructing tie

Standard tie length–57″
Width–3½″ wide at its widest point

Simple method of making tie without measurement
You can also draft a pattern yourself using an old tie your man no longer wears. Simply disassembled one, drafted a workable pattern from it.

5.9.6　Step by step procedure for making neck tie

Step 1
Draft the pattern by tracing the outline of the old tie pieces onto any kind of paper. I prefer the weight of craft paper. Ties are cut on the bias so pay attention to the grain line.

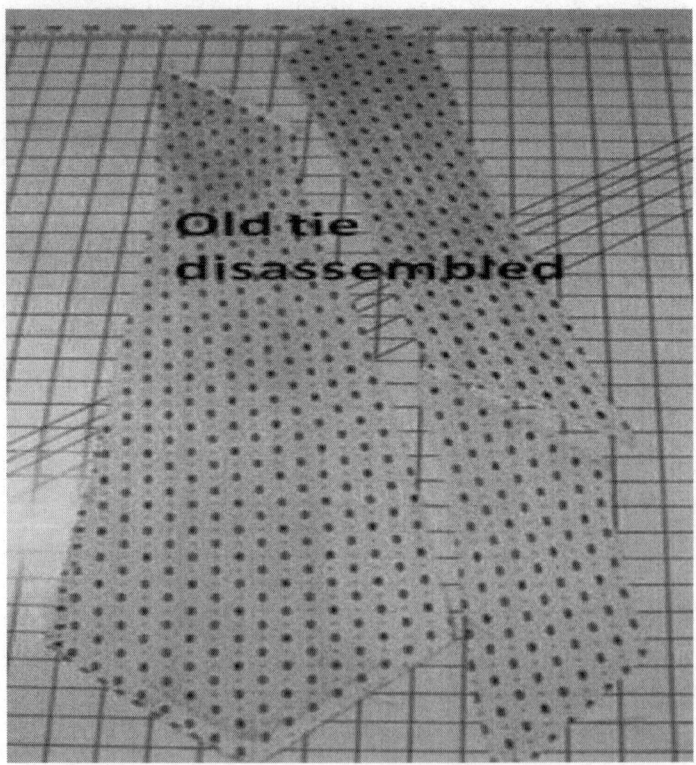

Step 2

Lay out the pattern pieces, pin in place and cut out with either shears or a rotary cutter.

Step 3

Join the three sections using ¼″ seam allowances, to form the base of the tie.

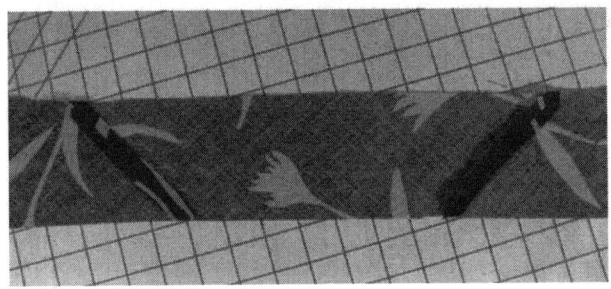

Step 4

Cut out the lining sections to form the tips.

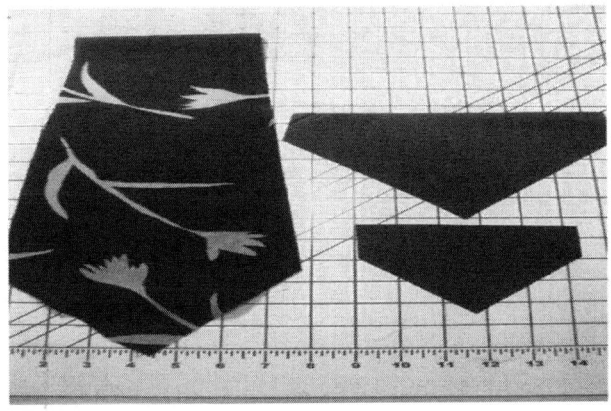

Step 5

Attach the lining to each tip end by first folding the tie ends in half and then stitching the tip closed roughly ¼″ from the tip end.

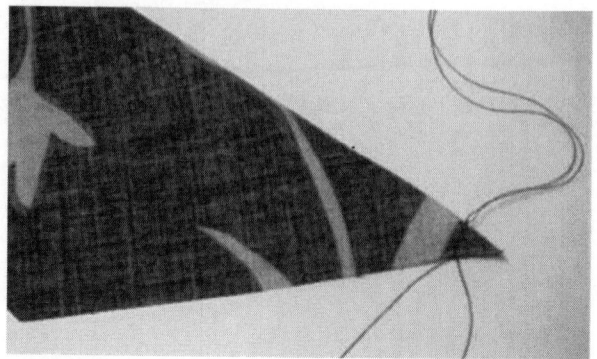

Step 6

Stitch the lining sides to the tie ends using ¼″ seam allowances and stitch to—but not over —the tip stitching.

Step 7

Carefully turn the tie point right side out and press in place. The lining edges should set back slightly from the tie point as shown above.

Step 8

Now, lay the interlining (I used the one from the old tie—so much easier) along the centre of the entire length of the tie. The end points should fit snugly into the tips at both ends.

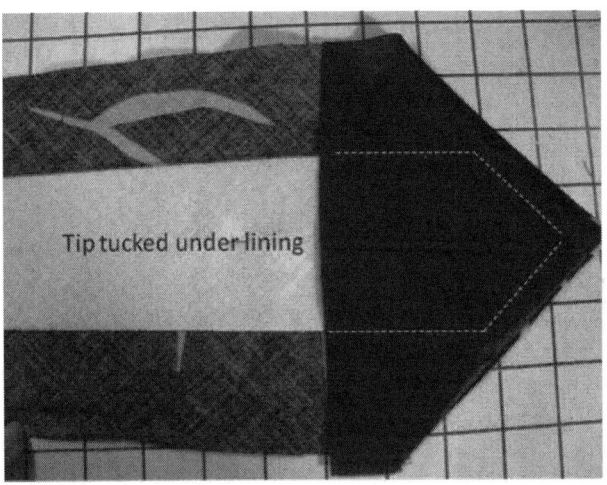

Tip tucked under lining

Step 9

Press the bottom edge up by about ⅜″ and press in place. Now fold the top edge down towards the centre and pin in place. Fold up the bottom edge towards the centre and pin that in place.

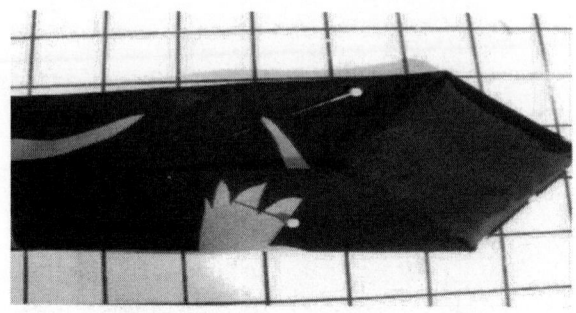

Step 10

Hand stitch the folded edge to the centre of the tie. Begin by sewing a bar tack and then stitch the length of the tie with a slip stitch. Stitch only through the interlining and not through to the base fabric. Finish with a bar tack at the opposite end.

Now, very lightly press to set stitches being careful not to flatten the edges.

5.10   Bags

5.10.1   Introduction

In the modern world, bags are ubiquitous, with many people routinely carrying a wide variety of them in the form of cloth or leather briefcases, handbags, and backpacks, and with bags made from more disposable materials such as paper or plastic being used for shopping, and to carry home groceries. Colours of the bags should either be in neutral colours that go with almost all outfits, or match with the colour of the outfit (or) shoes (or) can be in metallic shades for formal occasions. A bag may be closable by a zipper, snap fastener, etc., or simply by folding (e.g., in the case of a paper bag). Sometimes a money bag or travel bag has a lock. The bag likely predates the inflexible variant, the basket, and bags usually have the additional advantage over baskets of being foldable or otherwise compressible to smaller sizes. On the other hand, baskets, being made of a more rigid material, may better protect their contents. Fabric bags are a convenient and attractive. Hand bag accessory carried primarily by women and girls to hold such items as money credit cards and cosmetics.

5.10.2   Materials used for hand bags

Cotton: Cotton fabric comes in a variety of colours, prints, weight and textures. It is easily cut, quilted and embellished.

Faux fur: Faux fur has a nap and all the pattern pieces must be cut facing the same direction.

Leather and suede: Leather and suede is made from animal skin.

Linen: Linen fabrics have a loose weave and wrinkle easily unless fused with an interfacing or backed with batting and quilted. Linens can be pre-washed but it better to dry clean them.

Polyester and blends: Polyester and blends is very easy to care for. They can be pre-washed with very little shrinkage.

Rayon: Rayon fabric is versatile in texture, color and shine. It can have the drape and lustre of silk and is best dry cleaned.

Silk: Silk comes in different weights from sheer chiffon to slubby dupioni.

Tapestry: It is a textile with a loose weave but it can be fused with interfacing to stabilise.

Velvet and velveteen: Both fabrics have a directional nap and must be cut in the same direction.

Lining fabrics: There are various types of lining fabric in various colours and textures they add a professional look to the inside of the bag.

5.10.3   Different types of bags

All this bags are made out of fabrics.

1. Drawstring bags
2. Gift bags
3. Tote bags
4. Grocery bags.

5.10.3.1   Drawstring bags

Drawstring bags

Measurements

6.5 × 21″ piece of fabric, 22″ ribbon or cording.

Step by step procedure for making bag

Step 1

Turn the ends of your fabric under by a quarter of an inch and stitch in place for smooth edges.

Step 2

Turn the fabric right side up and stich the sides together with a ½″ seam, leaving a few inches open at the top.

Step 3

Iron the side seams open and secure them with a straight stitch.

Step 4

Create the drawstring opening by folding the top edge of your bag down by an inch and sewing all the way around. Make sure that you leave openings on either side of the bag.

Step 5

Feed one piece of your cord through the opening at the top of the bag, then feed it back through the other opening so that both ends of the cording are on the same side.

Step 6

Repeat with a second cord, feeding it in from the opposite side.

Step 7

Tie the ends of the string on each side together.

Step 8

Fill up your bag, and pull the strings tight.

5.10.3.2   Gift bags

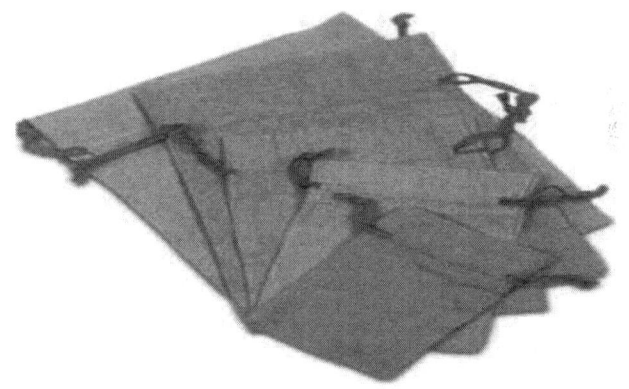

Gift bags

Measurements
Length–15″ Width–7″

Step by step procedure for making bag

Step 1

Cut your fabric into a rectangle that is twice the width you need.

Step 2

Fold it in half with the right sides together.

Step 3

Stitch down the sides and along the bottom with a ¼″ seam allowance, then turn the fabric right side out.

Step 4

Trim the top with pinking shears for a decorative edge.

Step 5

Fill your bag and tie it off with your favourite ribbon.

5.10.3.3   *Tote bags*

Tote bags

Measurements
1. Two 18 × 14″ pieces of canvas or cotton fabric, two 18 × 14″ pieces of interfacing, two 21″strips of canvas.

Step by step procedure for making bag

Step 1
To form the base of the fabric tote bag, trim a 2 × 2″ square from the bottom corners along the long side of all rectangles as well as the interfacing.

Step 2
Iron the interfacing to the wrong side of the outer fabric rectangle. With the right side together, sew ½″ seams along the sides and bottom of the piece. Leave the top corners and the two snipped corners unsewn.

Step 3
Pinch together the snipped corners so that the side and bottom seams are lined up in the centre and stitch with a ½″ seam allowance to create a flat bottom. Repeat with the inner liner fabric.

Step 4
Flip the inner liner right side out and place inside the outer fabric so that the right sides are together. Place the canvas straps between the inner and outer fabrics.

Step 5
With a ½″ seam allowance, stitch around the top of the tote, sewing the canvas straps in to the fabric. Be sure to leave a gap of approximately 4″.

Step 6

Turn the fabric tote bag right side out through the 4″ gap. Give your bag a final top stitch, and finished.

5.10.3.4   Grocery bags

Grocery bags

Step 1

Cut out two 18″ × 18″ squares of fabric and mark the top of each square (on the right side of the fabric) at 6½″ in from each side. This is where the handles will be attached. Then place the fabric so the two right sides are facing each other. Now cut a 2½″ square out of the bottom corner on each side.

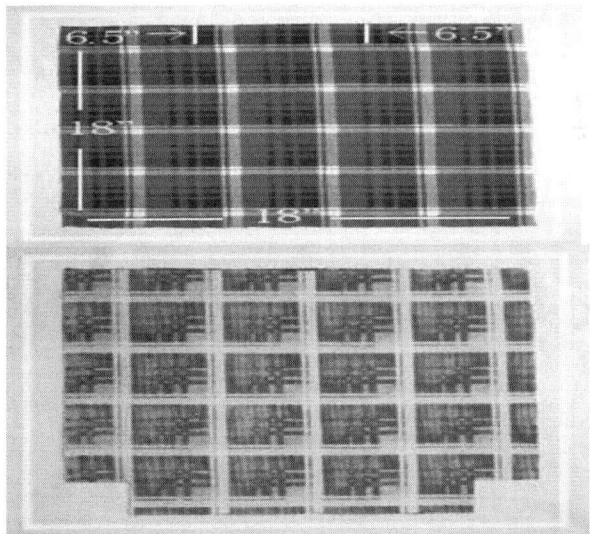

Step 2

With the right sides of the fabric facing each other, sew ¼″ from the edge, along both sides and across the bottom. Do not sew along the cut-out corners, leave those open.

Step 3

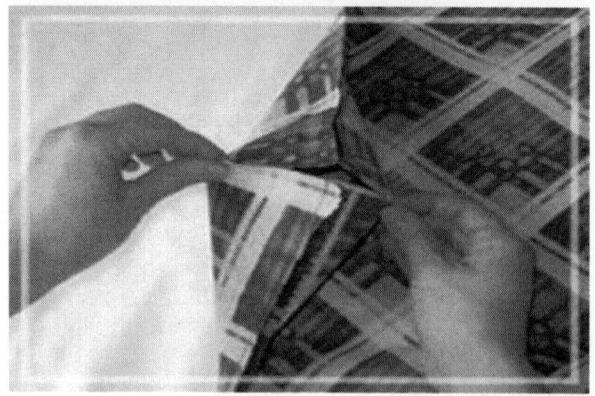

You will now be sewing the bottom corners closed. To get a corner ready to be sewn, open up the bag a little bit. Now on one side of the bag, pinch the fabric at each of the inner corner cuts. One corner from the front of the bag and one corner from the back. Line up the cut edges and sew about ½″ seam.

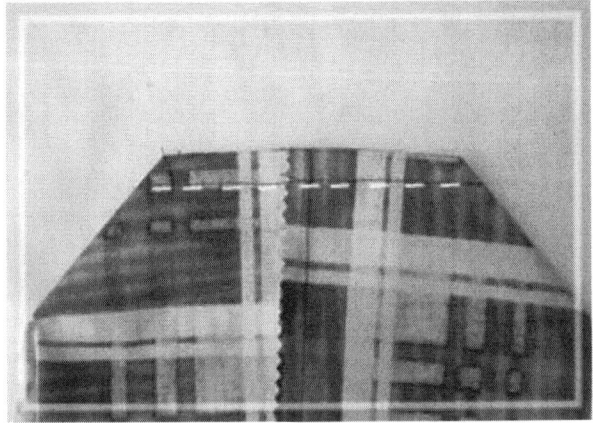

Step 4

Now cut two strips of fabric that are 3″ × 18″ for the handles. Fold the strip of fabric in half, length-wise, with the right sides facing out. Zig-zag stitch, or serge the open edge closed.

Step 5

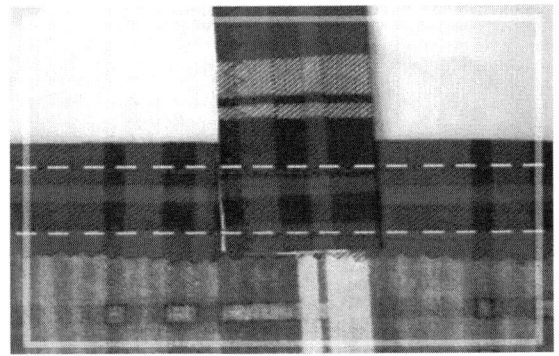

Fold over the top edge of the bag about 1″ or more, and sew a seam about ¾″ down from the top, around the whole bag. As you do this, sew down the ends of the handles at the marks you made in step 1. To reinforce the handles, backstitch over the end of the handle and then continue forward as you get to each handle. Be sure that the handles are not twisted when sewing them down.

After you've gone around the bag once, repeat this step again, at about 1/8″ down from the top edge of the bag, reinforcing the handles again.

5.11   Conclusion

Apparel and accessories are worn together, so one should not be considered independent of the other as one change, the other must also change to complement it to understand the directions of fashion, accessories are important.

5.12   Guide to gloves, caps, veils, belts, bows, ties and bags

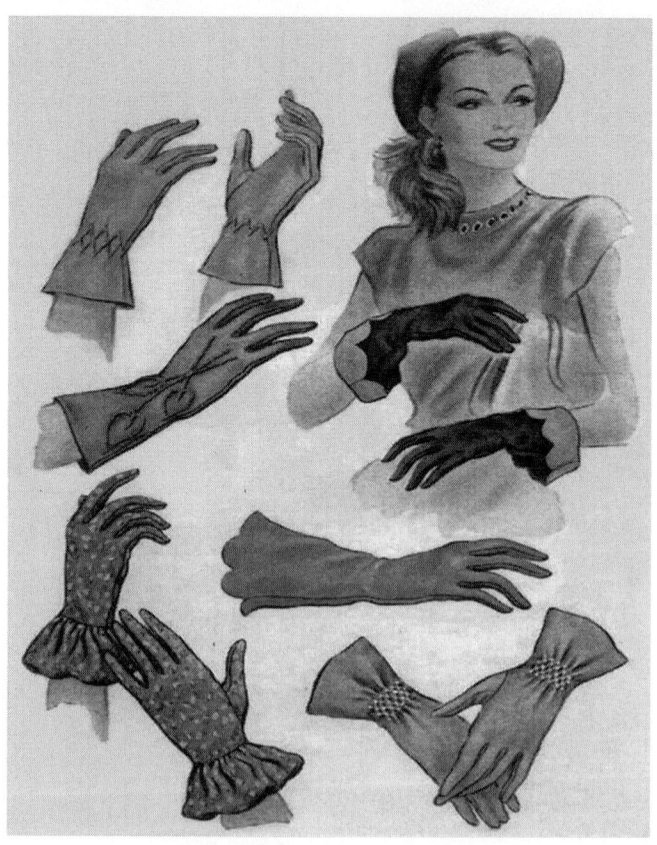

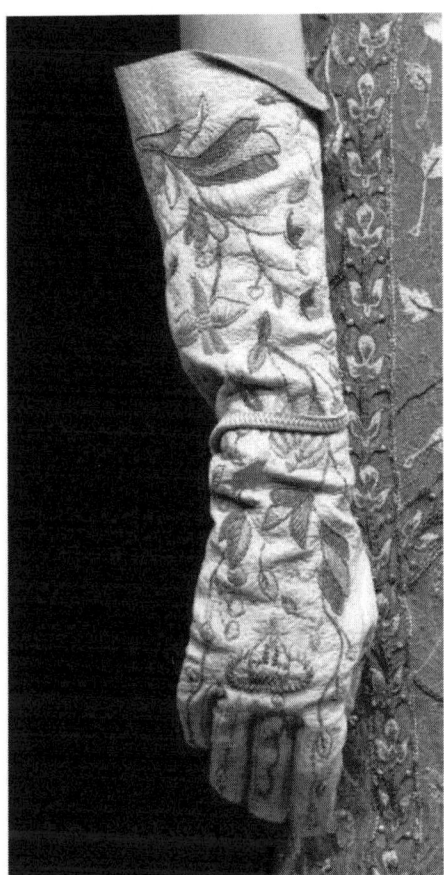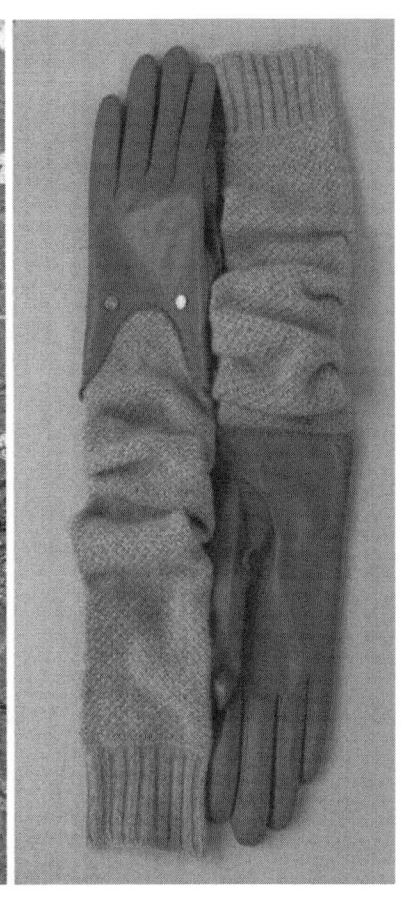

Guide to hats

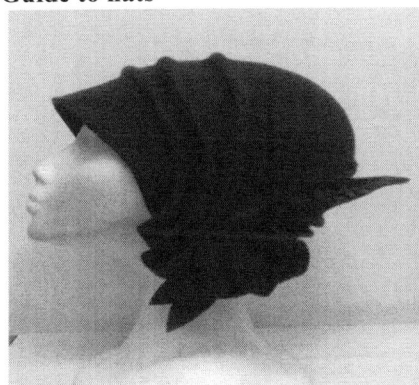

A Guide to Hats

Baseball

Beanie

Beret

Bowler (UK) / Derby (USA)

Bucket / Fisherman

Cloche

Cocktail

Cowboy

Deerstalker

Fascinator

Fedora

Gatsby

Homburg

Newsboy

Panama

Pillbox

Trapper

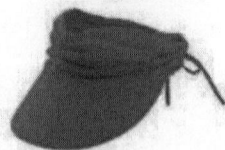

Visor

Guide to veils

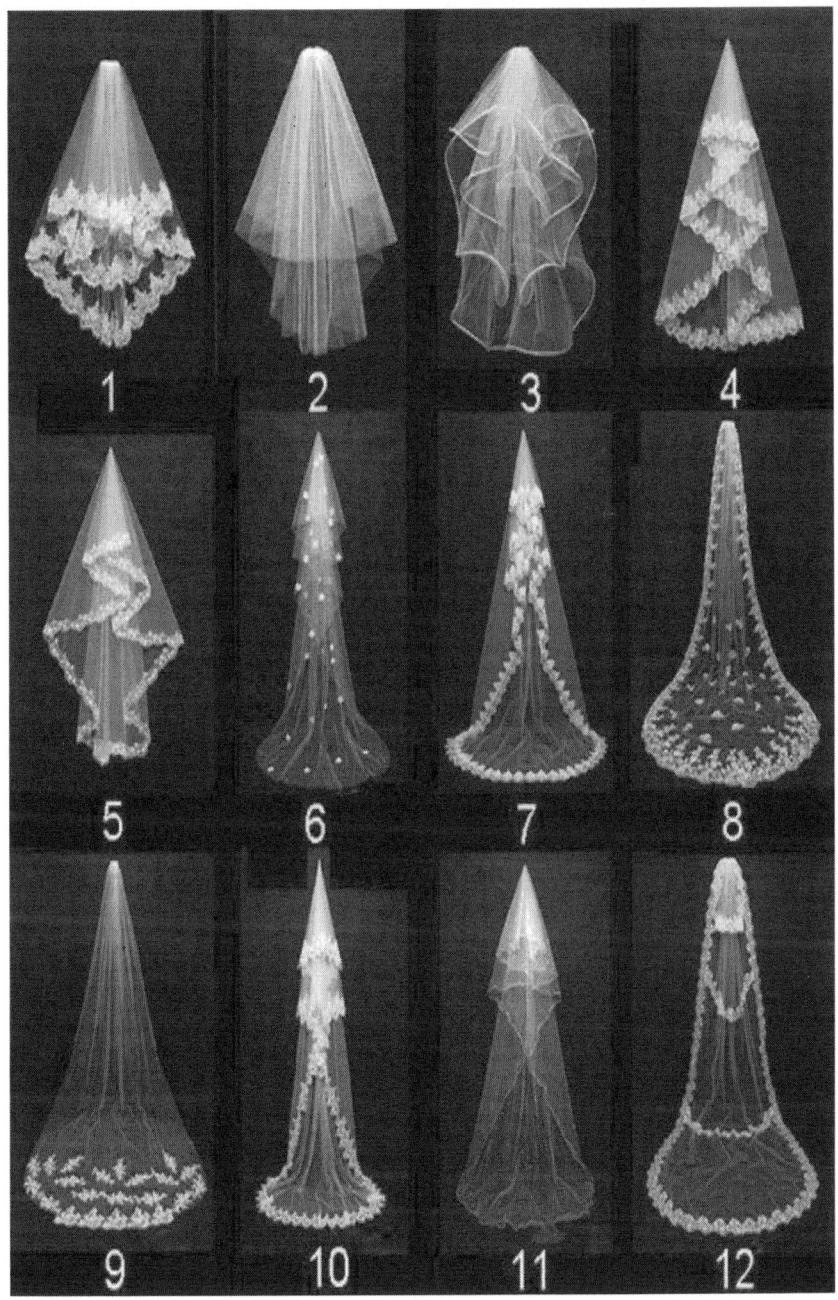

Guide to belts

Guide to bows

Guide to ties

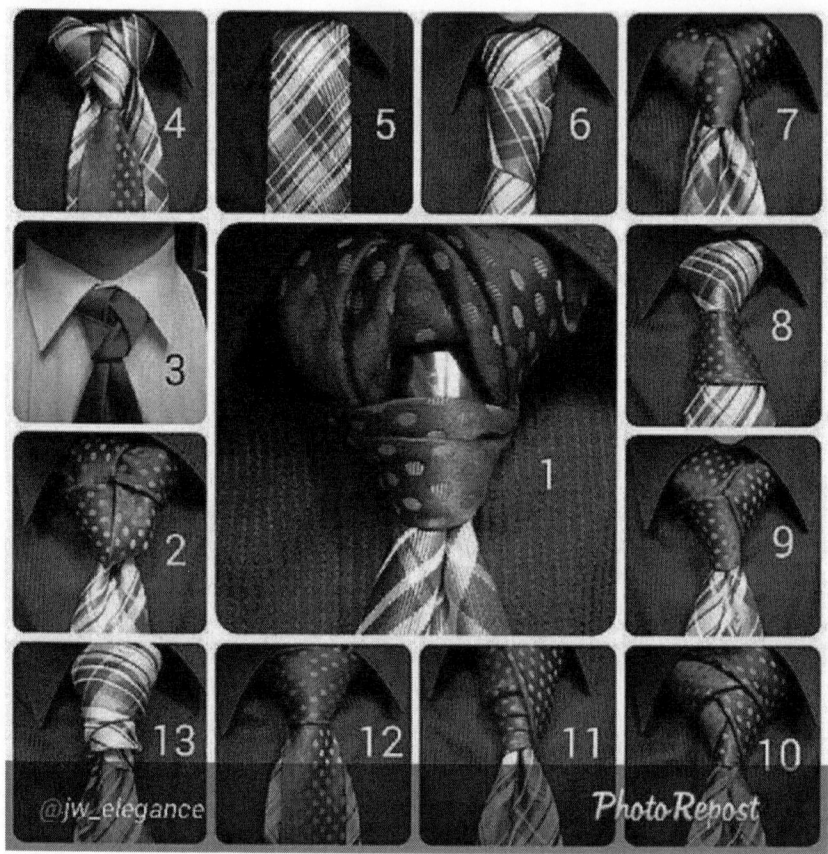

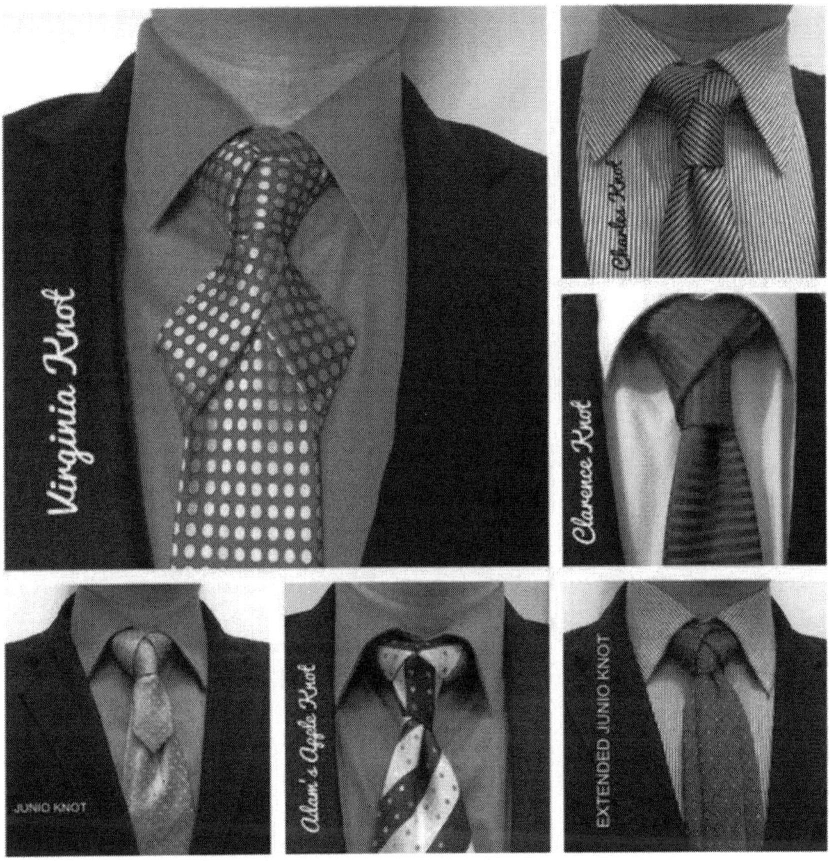

Virginia Knot

Charles Knot

Clarence Knot

JUNIO KNOT

Adam's Apple Knot

EXTENDED JUNIO KNOT

Guide to bags

5.13 References

1. Jennifer Alfano. The new secrets of style, designed by Bess Yoham. Melcher media, Instyle, ISBN-10:1-60320-082-7, ISBN 13:978-1-60320-082-0. p 166.

2. Amrit Bajaj, (2011). Creating sketching for embroidery. Sonali Publications, 4228/1, Ansari road, New Delhi-110002 (India), Arora offset press, Laxmi nagar, New Delhi-110092. ISBN-978-81-8411-353-2. p 147.

3. Charlotte mankey calasibetta Phyllis Tortora. Dictionary of fashion, Third edition, Illustrated by Bina abling, Fair child publications, Inc., ISBN: 81-8710-739-1, p 29.

4. Kathryn Mckelvey and Janine Munslow. Fashion forecasting. Wiley-Blackwell, A John Wiley and Sons, Ltd Publication, ISBN 978-1-4051-4004-1.

5. Farid Chenoune, (2005). Carried Away: All About Bags.

6. Bhargav, R., (2005). Design idea and accessories. First B. Jain, Publisher (P) Ltd, p 130–165.

7. Chenouna, F., (2004). Carried away, All about bags.

8. Rajkishore Nayak, Rajiv Padhye. Garment manufacturing technology. ISBN 978-1-78242-232-7, p 150–183.

9. Pauline Thomas, (2007). The Wearing of Hats Fashion History. Fashion-era.com. Retrieved 2011-07-02.

10. The social meanings of hats. University of Chicago Press. Retrieved 2011-07-02.

11. Insignia: The Way You Tell Who's Who in the Military. United States Department of Defence. Archived from the original on 2012-04-14. Retrieved 2011-07-02.

12. Posted by Don under accessory design.

13. www.google.com.

14. www.mademan.com/mm/how-make-socks.html.